Frontiers in Voltammetry

*Edited by Shashanka Rajendrachari,
Kiran Kenchappa Somashekharappa,
Sharath Peramenahalli Chikkegouda
and Shamanth Vasanth*

Published in London, United Kingdom

Frontiers in Voltammetry
http://dx.doi.org/10.5772/intechopen.102271
Edited by Shashanka Rajendrachari, Kiran Kenchappa Somashekharappa, Sharath Peramenahalli Chikkegouda and Shamanth Vasanth

Contributors
Mosaab Echabaane, Chérif Dridi, Harsha Devnani, Chetna Sharma, Subramaniam Rameshkumar, Panjaiyan Karthikeyan, Iman Danaee, Manogaran Obulichetty, Shashanka Rajendrachari, Rayappa Shrinivas Mahale, Sharath Peramenahalli Chikkegouda, Shamanth Vasanth, Kiran Kenchappa Somashekharappa, Rajni Bais

First published in London, United Kingdom, 2023 by IntechOpen
IntechOpen is the global imprint of INTECHOPEN LIMITED, registered in England and Wales, registration number: 11086078, 5 Princes Gate Court, London, SW7 2QJ, United Kingdom

British Library Cataloguing-in-Publication Data
A catalogue record for this book is available from the British Library

Additional hard and PDF copies can be obtained from orders@intechopen.com

Frontiers in Voltammetry
Edited by Shashanka Rajendrachari, Kiran Kenchappa Somashekharappa, Sharath Peramenahalli Chikkegouda and Shamanth Vasanth
p. cm.
Print ISBN 978-1-80356-446-3
Online ISBN 978-1-80356-447-0
eBook (PDF) ISBN 978-1-80356-448-7

Meet the editors

Dr. Shashanka Rajendrachari is currently working as an assistant professor in the Department of Metallurgical and Materials Engineering at Bartin University, Turkey. He completed an MSc in industrial chemistry and an MTech in nanoscience and technology at Kuvempu University, Karnataka, India, and obtained his Ph.D. from the Department of Metallurgical and Materials Engineering, National Institute of Technology, Rourkela, India. His research interests are nano-structured alloys, powder metallurgy, electrochemical sensors, corrosion, smart and high entropy alloys. He is the recipient of numerous national and international awards for his contribution to research. He has published more than 50 research articles, six books and five book chapters in various peer-reviewed international publications.

Dr. Kiran Kenchappa Somashekharappa is currently working in the Department of Chemistry, Government First Grade College, Shivamogga, India. He gained his Ph.D. from the Department of Nanotechnology at Visvesvaraya Technological University, CPGS-Bangalore Region, Muddenahalli. His research interests include the synthesis of nanomaterials with different morphology/methods and their characterizations and applications in the field of photocatalysis/catalysis, and sensors. Kiran has published research work in many peer-reviewed journals and has received Best Paper awards at several international conferences. He is a member of a number of US and Indian professional bodies including the Indian Science Congress Association, Institute of Chemists India, Society for Materials Chemistry, Materials Research Society and a Community Member of the American Chemical Society.

Dr. Sharath Peramenahalli Chikkegouda obtained his doctorate in metallurgical and materials engineering from NITK, Surathkal, India. He has 10 years of teaching/research experience and his current appointment is in the Metallurgical and Materials Engineering Department at Jain University. His research interests are metal forming, severe plastic deformation, composites and their characterization, additive manufacturing, electrochemical sensors and applications. He has published in more than 25 international journals and has chaired many conferences. He is a reviewer for many reputed international journals and received an Excellent Reviewer award from *ASM International - Journal of Materials Engineering and Performance*.

Dr. Shamanth Vasanth is currently with the School of Mechanical Engineering, REVA University, Bangalore, India. He worked as an engineer in the industry before his academic career at highly reputed Indian universities. He has educated dozens of engineers during his decade of teaching and research. He has more than 30 publications in the form of books, book chapters, and research articles. His research interests include additive manufacturing, powder metallurgy, heat treatment, advanced manufacturing systems and technologies.

Contents

Preface

Frontiers in Voltammetry focuses principally on the different applications of various modern voltammetric techniques. The five chapters are all written in simple easy-to-understand language.

Chapter 1 mainly discusses the successful use of electrochemical impedance spectroscopy (EIS) to determine the molecular weight of compounds. The authors discuss the advantages of this technique over such conventional ones as Ostwald and Walker's method. Generally, the measurement of the vapor pressure of pure solvents in solutions is difficult and lengthy in Ostwald and Walker's method, while EIS is a simple, robust, accurate and non-destructive method of determining the molecular weight of polymers and proteins soluble in protic solvents.

Chapter 2 reports the use of carbon dot and chitosan nanocomposite-based sensors to detect copper ions using EIS. Prepared carbon dots and the modified glassy carbon electrode are characterized by transmission electron microscope, scanning electron microscope (SEM), Fourier-transform infrared spectroscopy (FTIR), X-Ray diffraction, and UV-visible spectroscopy. From their discussion of the stability, selectivity, specificity, sensitivity, dynamic range, and detection limits, the authors conclude that fabricated electrodes are economic, simple, and highly sensitive, and give better current response during copper ion detection.

Chapter 3 considers how the many applications of electrochemical sensors could enhance the sensitivity, selectivity, robustness, accuracy, and anti-fouling properties of the electrodes in various fields including healthcare, biomedical, pharmaceuticals, and agriculture. These types of electrochemical sensors are much sought after to meet global demand in various sectors [1–11]. Nanomaterials are a popular choice as a modifier of electrode sensors because of their possible immobilization of molecules, catalysis in electrosynthesis, facilitation of electron transfer between electrodes and biomolecules, and labeling of biomolecules.

Chapter 4, written by the editors, discusses the use of various electrochemical sensors to detect toxic heavy metals like arsenic, cadmium, and lead heavy metals which can cause malfunction of the gastrointestinal, nervous, kidney, and immune systems, birth defects, skin lesions, and cancer. The main source of heavy metal pollution is geological activities or agricultural and industrial wastes (**Figure 1**) [12, 13]. Cyclic voltammetry has proved to be one of the most popular methods of detecting various heavy metals using different modified carbon paste electrodes, carbon screen-printed electrodes, and glassy carbon electrodes. The authors also discuss the advantages of cyclic voltammetry over other non voltammetric methods for detecting heavy metals.

The final chapter discusses the importance of developing toxic organic and inorganic sensors for use in a wide variety of fields, including food sensors, drug and pharmaceutical sensors, biosensors, agricultural residue sensors, energy sensors, and forensic and environmental sensors, to avoid environmental health risks. The authors explain that electroanalytical techniques are better than spectrometric or chromatographic techniques in sensing applications. Among the electroanalytical techniques described are linear

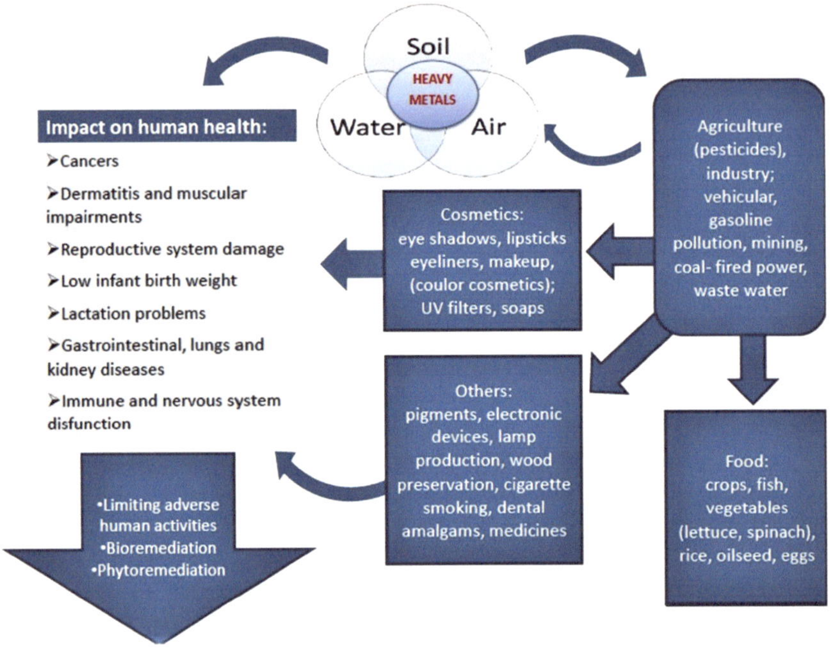

Figure 1.
Pictorial representation of heavy metal, their effect on human health, and their mechanism [12].

sweep voltammetry (LSV), cyclic voltammetry (CV), pulse voltammetry (PV), stripping voltammetry (SV), and chronoamperometry.

Dr. Shashanka Rajendrachari
Assistant Professor,
Department of Metallurgical and Materials Engineering,
Bartin University,
Bartin, Turkey

Kiran Kenchappa Somashekharappa
Department of Chemistry,
Government First Grade College,
Shivamogga, Karnataka, India

Sharath Peramenahalli Chikkegouda
Department of Metallurgical and Materials Engineering JAIN Deemed
to be University,
Bengaluru, Karnataka, India

Shamanth Vasanth
School of Mechanical Engineering,
REVA University,
Bengaluru, Karnataka, India

References

[1] Mahale RS, Vasanth S, Krishna H, Shashanka R, Sharath PC, Sreekanth NV. Electrochemical sensor applications of nanoparticle modified carbon paste electrodes to detect various neurotransmitters: A review. Applied Mechanics and Materials. 2022;**908**:69-88

[2] Pavitra V, Praveen BM, Nagaraju G et al. Energy storage, photocatalytic and electrochemical nitrite sensing of ultrasound-assisted stable Ta_2O_5 nanoparticles. Topics in Catalaysis. 2022. DOI: 10.1007/s11244-021-01553-7

[3] Shashanka R. Investigation of optical and thermal properties of CuO and ZnO nanoparticles prepared by *Crocus sativus* (Saffron) flower extract. Journal of the Iranian Chemical Society. 2021;**18**:415-427

[4] Shashanka R, Karaoglanli AC, Ceylan Y, Uzun O. A fast and robust approach for the green synthesis of spherical magnetite (Fe_3O_4) nanoparticles by *Tilia tomentosa* (Ihlamur) leaves and its antibacterial studies. Pharmaceutical Sciences. 2020;**26**:175-183

[5] Shashanka R, Dileep R. Chapter 1: Functionalized nanomaterial-based electrochemical sensors: A sensitive sensor platform. In: Hussain C, Manjunatha JG, editors. Functionalized Nanomaterial-Based Electrochemical Sensors. Amsterdam: Elsevier, Woodhead Publishing. 2022; Paperback ISBN: 9780128237885, eBook ISBN: 9780128241851

[6] Kiran KS, Shashanka R. Sustainable development information management of carbon nanomaterials-based sensors. In: Hussain C, Manjunatha JG, editors. Carbon Nanomaterials-Based Sensors. Amsterdam: Elsevier; 2022. ISBN: 978-0-323-91174-0. DOI: 10.1016/B978-0-323-91174-0.00001-9

[7] Shashanka R, Kumara Swamy BE. Simultaneous electrogeneration and electrodeposition of copper oxide nanoparticles on glassy carbon electrode and its sensor application. SN Applied Sciences. 2020;**2**(5):956

[8] Shashanka R, Kumara Swamy BE. Biosynthesis of silver nanoparticles using leaves of *Acacia melanoxylon* and its application as dopamine and hydrogen peroxide sensors. Physical Chemistry Research. 2020;**8**(1):1-18

[9] Rajendrachari S, Adimule V, Gulen M, Khosravi F, Somashekharappa KK. Synthesis and characterization of high entropy alloy 23Fe-21Cr-18Ni-20Ti-18Mn for electrochemical sensor applications. Materials. 2022;**15**:7591

[10] Shashanka R. Carbon composite voltammetric sensors for food quality assessment. In: Manjunatha JG, editor. Electrochemical Sensors Based on Carbon Composite Materials. Bristol, UK: IOP Publishing; 2022. ISBN: 978-0-7503-5127-0. DOI: 10.1088/978-0-7503-5127-0ch8

[11] Mahale RS, Shashanka R, Shamanth V, Vinaykumar R. Voltammetric determination of various food azo dyes using different modified carbon paste electrodes. Biointerface Research in Applied Chemistry. 2022;**12**(4):4557–4566

[12] Witkowska D, Słowik J, Chilicka K. Heavy metals and human health: Possible exposure pathways and the competition for protein binding sites.

Molecules. 2021;**26**:6060. DOI: 10.3390/molecules26196060

[13] Balali-Mood M, Naseri K, Tahergorabi Z, Khazdair MR, Sadeghi M. Toxic mechanisms of five heavy metals: Mercury, lead, chromium, cadmium, and arsenic. Frontiers in Pharmacology. 2021;**12**:643972. DOI: 10.3389/fphar.2021.643972

Section 1

Applications of Electrochemical Impedance Spectroscopy (EIS) in Various Fields

Chapter 1

A New Method for Determination of Molecular Weight of Compounds Soluble in Protic Solvents by Electrochemical Impedance Spectroscopy

Subramaniam Rameshkumar, Panjaiyan Karthikeyan, Iman Danaee and Manogaran Obulichetty

Abstract

This chapter deals with a new method for determining the molecular weight of chemical substances soluble in protic solvents. One of the well-known methods for the determination of molecular weight of a substance, based on one of the colligative properties, is Ostwald and Walker's method, which depends on relative lowering of vapor pressure of solvent. In this paper we proposed a new method for determining the molecular mass of the substances that are soluble in protic solvents such as water, methanol and ethanol employing electrochemical impedance spectroscopy (EIS) technique and Raoult's law. The moisture and vapor pressure dependent proton conductivity of some organic compounds and metal-organic frame works (MOFs) can be utilized to find the molecular mass of solutes soluble in protic solvents. This property is considered as key for determination of molecular weight of chemical substances using EIS and is simpler than Ostwald and Walker's method. This method is a non-destructive and also useful to determine the molecular weight of polymers and proteins soluble in protic solvents.

Keywords: impedance, molecular weight, vapor pressure, Raoult's law

1. Introduction

Electrochemical impedance spectroscopy (EIS) finds a special place among the various electrochemical techniques. It is a powerful tool for analyzing the interfaces formed in the heterogeneous systems. EIS supplies a large amount of informations, though it may not provide all the answers. EIS uses tools developed in electrical engineering and describes the behavior of the systems under study in terms of an equivalent circuit consisting of the circuit elements resistors, capacitors, Warburg impedance etc. The mathematical foundations of EIS were dealt by Heariside [1], through which it is possible to solve the integrodifferential equations appearing in the solutions of electrical circuits by converting them into a system of algebraic equations. The main advantage of EIS is the fact that it is based on the linear time invariant (LTI) theory and validity of the data may be verified using Kramers-Kronig integral transforms. Nernst was the first person who described the chemical applications of impedance spectroscopy through his work [2], followed by many others including those applications to the distribution of relaxation time constants by Cole and Cole [3] and Davidson and Cole [4]. The impedance of mass transfer was explained by Warburg using so called Warburg impedance, which extended EIS to apply for redox reactions [5]. With further development in the understanding of EIS, the structure of double layer in the absence and presence of adsorbed species was studied initially at dropping mercury electrode and then at solid electrodes using AC bridge. The analysis of electrochemical reactions using the electro analog circuit was introduced by Dolin and Ershler [6] and Randles [7, 8], where the age of electrical analog began [9] and continues up till now. The fundamental aspects of EIS give the idea to validate the data and to model the processes limited by diffusion, electrode kinetics and adsorption on different types of electrode geometries. The availability of modern instrumentation to obtain impedance data as well as computer programs to interpret the results have made this technique popular. Now a days, EIS finds applications in corrosion, biosensors, battery development, fuel cell development, drug cell membrane interaction [10], paint characterization, sensor development, polymers etc.

The development of organic proton conducting materials to substitute the per fluorinated polymers such as nafion is an important area of research in the field of fuel cell technologies [11–15]. Recent research in this field emerged some coordination polymers and metal-organic frame works (MOFs) for their proton transport capabilities, though the literature on these materials in other domains such as magnetism, catalysis, inclusion phenomenon and in supramolecular chemistry are quite extensive [16–23]. Under high humidity conditions or with water channels, MOFs show high degree of proton conductivities, comparable to nafion [24–28]. The proton conducting ability of these materials primarily depend on the existence of charge carrying molecular or ionic species such as H_3O^+, OH^-, or NH_4^+ Lewis acidic moieties. These molecular or ionic species with the complex network of hydrogen bonds or their arrangements with the water molecules play a vital role for proton conductivity. These characteristics are considered to be the important structural requirements of the proton conducting polymers and MOFs [29–33]. With the above-mentioned features, several new MOFs that exhibit Nafion-like proton conducting characters, under humidified and ambient temperature conditions have been reported [34]. The proton conducting ability of these materials was evaluated using EIS [34–36] which depends on humidity levels of the surroundings.

In this chapter, through an innovative approach we would like to propose a new method for the determination of molecular weight of compounds which are soluble in water and other protic solvents, using proton conducting ability of organic compounds or MOFs by EIS technique.

2. Electrochemical impedance spectroscopy—a tool to evaluate the proton conductivity of solid materials theory and discussion

Now a days proton conducting solid state electrolytes gained a considerable attention owing to their application in fuel cells1, electrochromic devices [37], humidity sensors [38] and gas separators [39]. Nafion, a perfluorinated polymer with sulphonic acid is used as solid-state proton conductor under hydrated conditions [40]. This polymer has many disadvantages, under high temperature (>80°C) the water clusters are lost in the pores of Nafion. At low humidity levels water clusters in the Nafion pores are low. Both factors reduce the conductivity of Nafion [40]. In addition to these drawbacks, high cost, high fuel cross over, non eco-friendly synthesis and variation of conductivity with degree of sulphonation limit its applications. These limitations made the researchers to look for alternate solid state proton conducting materials. Many coordination compounds and MOFs having proton conducting ability were not explored for their proton conductivities. Recent reports on proton conducting nature of MOFs promise their enhanced proton conductivity at high temperature and low fuel cross over.

It was reported that the organic compounds with heteroatoms incorporated within the pores of MOFs as guests with controlled loading. The amphoteric heterocyclic moiety has shown electrolytic conductance through proton. This system has shown an improved proton conductance even at higher temperatures and precisely the theoretical open circuit potential of hydrogen-oxygen fuel cell. Only few compounds have been studied for their proton conductivities. Their proton conducting properties can further be extended to calculate the molecular weight of solutes dissolved in the protic solvents and the vapor pressure of the protic solvents at any temperature knowing the normal boiling point. Conversely, this property can also be used for the calculation of molar enthalpy of vaporization of protic solvents. These applications have not been verified for these compounds. This method could be an alternate for Ostwald and Walker Method for determining molecular weight of substances and a useful technique for knowing the molecular weight of polymers soluble and an unknown substance in protic solvent. The proposed method is based on EIS technique.

EIS is a routine method for characterization of various electrical properties of different types of materials and the interfaces formed by the materials with electronically conducting materials. EIS is a simple non-destroying technique, where a system study is perturbed by an AC sinusoidal voltage of small amplitude at different applied frequencies and the resulting current varying with the applied frequency is used to extract the required kinetic informations.

2.1 Fundamentals of electrochemical impedance spectroscopy

The resistance of a system under study comes from the hindrance offered for the flow of electrical current through its circuit elements. The resistance, R, of the material is defined by Ohm's law, in terms of voltage, E and current I ratio:

$$R = \frac{E}{I} \tag{1}$$

Similar to resistance, the term impedance measures the ability of a material to resist the flow of charges in definite direction. However, impedance differs from resistance in two main aspects. First, it is associated with alternating current (AC); second, it is usually mentioned at a particular frequency.

Measurement of impedance is done by applying a sinusoidally varying an AC potential to the system under study and measuring the current through the system.

The response to the applied sinusoidally varying potential at a frequency is a sinusoidally varying current at the same frequency with shift in phase. The ratio of applied sinusoidally varying potential to the resulting sinusoidally varying current at the same frequency is called impedance, using which the information on conductivity of the system can be obtained.

The input or excitation signal, i.e. sinusoidally varying potential at a frequency can be expressed as a function of time as follows:

$$E(t) = E_0 \sin(\omega t) \tag{2}$$

Where,

$E(t)$ refers to the potential at time t, E_0 is the amplitude of the applied AC potential and ω is the angular frequency given by the expression $\omega = 2\pi f$ (f = applied frequency in Hz)

In a linear system the excitation of voltage at a frequency 'ω' provides a current 'I' at the same frequency. This generated current is different in amplitude and phase from voltage as given below:

$$I(t) = I_0 \sin(\omega t + \varphi) \tag{3}$$

In a non-linear system I-V relation gives a distorted response which is not purely sinusoidal, but it is still periodic [41].

An analogous expression to Ohm's law allows us to calculate the impedance of the system as:

$$Z = \frac{E(t)}{I(t)} = \frac{E_0 \sin(\omega t)}{I_0 \sin(\omega t + \varphi)} = Z_0 \frac{\sin(\omega t)}{\sin(\omega t + \varphi)} \tag{4}$$

The impedance is therefore expressed in terms of a magnitude, Z_0, and a phase shift (φ). Using Euler's relationship, the impedance is then represented as a complex number

$$Z = \frac{E(t)}{I(t)} = \frac{E_0 \sin(\omega t)}{I_0 \sin(\omega t + \varphi)} = Z_0(\cos\varphi + j\sin\varphi) \tag{5}$$

Accordingly, the impedance is a vector quantity since it has a magnitude, Z_0, and direction given by phase shift, φ.

When we apply the above concepts to a pure resistor, for which the phase angle (φ) is zero,

$$E(t) = E_0 \sin(\omega t) \tag{6}$$

$$I(t) = I_0 \sin(\omega t) \tag{7}$$

$$R = \frac{E(t)}{I(t)} = \frac{E_0 \sin(\omega t)}{I_0 \sin(\omega t)} = \frac{E_0}{I_0} \tag{8}$$

When the concept is applied for a capacitor, where, $\varphi = \frac{\pi}{2}$

Charge in the capacitor,

$$q = CE \tag{9}$$

Current,

$$I = \frac{dq}{dt} = C = \frac{dE}{dt} \tag{10}$$

$$I = CE_0\omega \cos(\omega t) \tag{11}$$

$$I = \frac{E_0}{X_C} \sin\left(\omega t + \frac{\pi}{2}\right) \tag{12}$$

Where X_C is capacitive reactance, i.e. the resistance offered by a pure capacitor for the alternate current flows across it. Its value depends on frequency of AC excitation voltage and is given as:

$$X_C = \frac{1}{\omega C} \tag{13}$$

Where, '*C*' is capacitance of the pure capacitor (in F).

Since the current leads the voltage by a phase angle of $\frac{\pi}{2}$, the capacitive reactance is taken along the ordinate (*Y*-axis) and is given as imaginary component.

$$\text{i.e.} \quad -iX_C = \frac{-i}{\omega C} = \frac{E(t)}{I(t)} \tag{14}$$

Where, $i = \sqrt{-1}$, an imaginary number.

When a capacitor and a resistor are in series connection, the excitation voltage is given as,

$$E = E_R + E_C \tag{15}$$

$$E = IR - iX_C I \tag{16}$$

$$\frac{E}{I} = R - iX_c \tag{17}$$

Since the value of X_C depends on frequency, the total resistance in this series depends on the frequency of excitation voltage and is called impedance (*Z*).

$$Z = R - iX_C \tag{18}$$

In general, the impedance can be represented as,

$$Z(\omega) = Z_{Re} - iZ_{Im} \tag{19}$$

Where, Z_{Re} = real part of impedance,

Z_{Im} = imaginary part of impedance.

The magnitude of impedance is given as,

$$|Z| = \sqrt{(Z_{Re})^2 + (Z_{Im})^2} \tag{20}$$

$$|Z| = \sqrt{R^2 + \frac{1}{\omega^2 C^2}} \tag{21}$$

The phase angle φ is given as,

$$\varphi = \tan^{-1}(Z_{\text{Im}}/Z_{\text{Re}}) \tag{22}$$

$$\varphi = \tan^{-1}(X_C/R) \tag{23}$$

$$\varphi = \tan^{-1}(1/\omega CR) \tag{24}$$

It is well known that impedance is a specific form of the transfer function of the system [41]. If $\bar{I}(s)$ and $\bar{E}(s)$ are the Laplace transforms of the sinusoidal current and voltage respectively, the transfer function [41] is given as,

$$\text{TF}(s) \cong \frac{\bar{E}(s)}{\bar{I}(s)} \simeq \frac{E_0}{I_0}\left(\cos\varphi + \frac{s}{\omega}\sin\varphi\right) \tag{25}$$

Were, s is the Laplace complex variable or frequency,

$$s = i\omega \tag{26}$$

Now solutions to the Eq. (26) are confined to the frequency domain.

$$Z \cong \text{TF}(i\omega) = \frac{E_0}{I_0}(\cos\varphi + i\sin\varphi) \tag{27}$$

2.2 Variation of frequency

When we consider a homogeneous system, it is represented by a conductance (G) or resistor (R) in parallel connection with a capacitor (C) as shown in **Figure 1**.

The corresponding Nyquist and Bode (phase and impedance) plots for the above equivalent circuit are presented in the **Figures 2** and **3** respectively.

By rule the net impedance of this homogeneous system, from the above equivalent circuit, is given as,

$$\frac{1}{Z} = \frac{1}{R} - \frac{1}{iX_C} \tag{28}$$

$$\frac{1}{Z} = \frac{1}{R} + \frac{i}{X_C} \tag{29}$$

$$\frac{1}{Z} = G + i\omega C \tag{30}$$

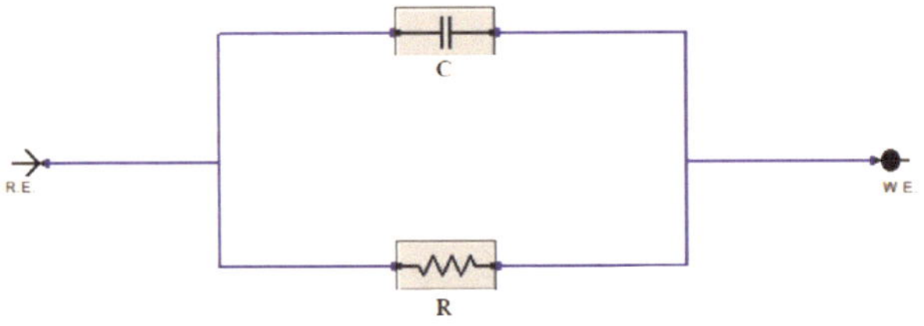

Figure 1.
An equivalent circuit for a homogeneous system.

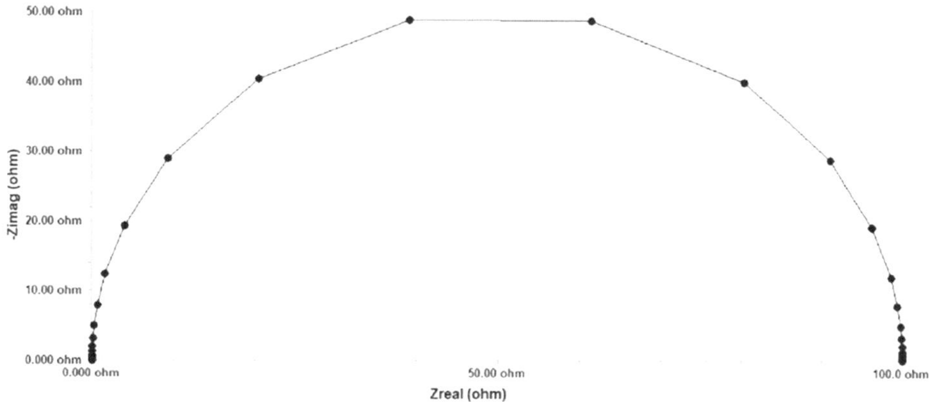

Figure 2.
Nyquist plot for the equivalent circuit proposed for homogeneous system (arbitrarily assumed value for R *is 100 Ω and* C *is 5 × 10^−6 F).*

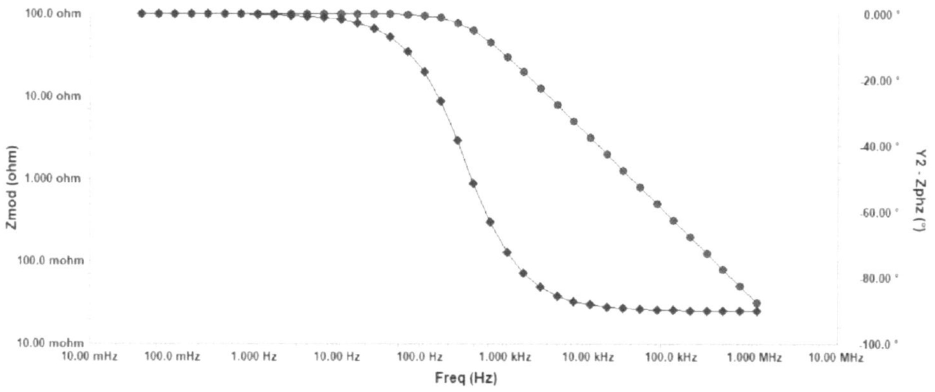

Figure 3.
Bode impedance and phase angle plots for the equivalent circuit proposed for homogeneous system (arbitrarily assumed value for R *is 100 Ω and* C *is 5 × 10−^6 F).*

$$Z = \frac{1}{G + i\omega C} \tag{31}$$

Conductance (G) and capacitance (C) of the homogeneous material describe its ability to conduct and store electric charge respectively. If the homogeneous material is considered as a slab of cross-sectional area 'A' and thickness 'l', these properties are given by the following expressions,

$$G = \sigma \frac{A}{l} \tag{32}$$

$$C = \varepsilon \frac{A}{l} \tag{33}$$

Where σ and ε are conductivity and permittivity of the homogeneous system respectively.

From Eqs. (27) and (31),

$$\frac{1}{G + i\omega C} = \frac{E_0}{I_0}(\cos\varphi + i\sin\varphi) \tag{34}$$

$$\frac{1}{G + i\omega C} = |Z|(\cos\varphi + i\sin\varphi) \tag{35}$$

$$G + i\omega C = \frac{1}{|Z|}(\cos\varphi - i\sin\varphi) \tag{36}$$

Therefore,

$$G = \frac{1}{|Z|}\cos\varphi \tag{37}$$

$$C = -\frac{1}{|Z|\omega}\sin\varphi \tag{38}$$

Using the above equations, we can measure the thickness of the homogeneous material.

Inspection of the above equation reveals that the impedance of the homogeneous material is dispersed with frequency and the dispersion is more pronounced for the frequencies greater than G/C,

$$\omega = \frac{G}{C} \tag{39}$$

Giving the condition at which, the impedance measurement provides the most accurate estimates of both properties. This characteristic frequency of the homogeneous system is called 'natural frequency' of the system.

A heterogeneous system can be considered to consist of a number of different materials or slabs sandwiched together. The total impedance of such a system is given as,

$$Z_{\mathrm{N}}(\omega) = \sum_{n=1}^{N} \frac{1}{\mathrm{G}_n + i\omega C_{\mathrm{n}}} \tag{40}$$

Each slab in the heterogeneous system has its own natural frequency defined by the expression $\omega = \frac{G}{C}$. If the magnitudes of impedance of corresponding N slabs are sufficiently different, then these slabs can be easily identified within the combined dispersions immediately.

For example, using the last expression, the total impedance of a heterogeneous system consisting of two different materials or slabs (P and M), the impedance dispersion can be written as,

$$Z_{\mathrm{N}}(\omega) = \frac{1}{G_{\mathrm{P}} + \mathrm{i}\omega C_{\mathrm{P}}} + \frac{1}{G_{\mathrm{M}} + i\omega C_{\mathrm{M}}} \tag{41}$$

Dispersion of conductance (G) and capacitance (C) with frequency are obtained by simplifying this equation as,

$$G = \frac{G_M G_P(G_M + G_P) + \omega^2(C_M^2 G_P + C_P^2 G_M)}{(G_P + C_M)^2 + \omega^2(C_P + C_M)^2} \quad (42)$$

$$C = \frac{C_M G_P^2 + C_P G_M^2 + (\omega^2 C_M C_P)(C_P + C_M)}{(G_P + C_M)^2 + \omega^2(C_P + C_M)^2} \quad (43)$$

The solid-state proton conducting electrolytes can be synthesized by simple solution crystallization method. The Nyquist plots recorded will not be straight forward to give the electrical properties of solid-state proton conductors. Equivalent circuits are used to derive the electrical properties of solid-state electrolytes. The equivalent circuit that fits well with the Nyquist plots will also provide the nature of arrangement of different dielectric slabs inside the soild-state protonic conductor. The mechanism for proton conduction could be derived from the temperature studies.

3. Determination of molecular mass using the electrochemical impedance spectra of solid-state proton conductors

Vapor pressure is a characteristic property of a substance in the condensed phase at constant temperature. It is the pressure exerted by the vapor molecules of a substance on the surface of the condensed phase of the same substance, when the vapor molecules are in equilibrium with its condensed phase at constant temperature. The vapor pressure of a liquid at constant temperature is constant and increases with temperature. At one temperature the vapor pressure of the liquid becomes equal to atmospheric pressure at which the liquid boils and temperature is called boiling point of the liquid. At the same temperature the vapor pressure of a liquid decreases when a non-volatile solute is dissolved in it which is called lowering of vapor pressure of the solvent. The proton conducting ability of coordination polymers and MOFs depends on the humidity levels or water vapor level in an environment. Greater the humidity level greater will be the proton conductivity. The proton conductivity of a coordination polymer or MOF is measured from the charge transfer resistance (R_{ct}) value obtained from EIS [34–36]. The charge transfer resistance measured using EIS is inversely proportional to proton conducting of the CPs or MOFs, which in turn directly proportional to humidity level or moisture level in the atmosphere.

Therefore, the charge transfer resistance (R_{ct}) of CPs or MOFs in the presence of known volume of pure water or protic solvents, measured in a closed container,

$$R_{ct} \ \alpha \ \frac{1}{P^0} \quad (44)$$

Where P^0 is vapor pressure of pure protic solvent.

The charge transfer resistance measured after dissolving a known weight of solute in the same volume of water or protic solvents in the closed container at same temperature,

$$R'_{ct} \ \alpha \ \frac{1}{P} \quad (45)$$

Where P is vapor pressure of solvent in solution
From Raoult's law

$$P = P^0 X_1 \tag{46}$$

or

$$X_1 = \frac{P}{P^0} \tag{47}$$

Where X_1 is mole fraction of protic solvent.
Therefore

$$X_1 = \frac{R_{ct}}{R'_{ct}} \tag{48}$$

$$\frac{n_1}{n_1 + n_2} = \frac{R_{ct}}{R'_{ct}} \tag{49}$$

Where $n_1$and n_2 are number of moles of solvent and solute respectively.

$$M_2 = \frac{W_2 M_1 R_{ct}}{W_1 \left(R'_{ct} - R_{ct}\right)} \tag{50}$$

Using Eq. (50) the molecular weight of the solute can be calculated.

This method is also applicable for determining molecular weight of the substances which are insoluble in water but soluble in methanol or ethanol, since some CPs or MOFs showing reversible proton conductivity in CH_3CH_2OH are also reported [35, 36].

4. Conclusion

The measurement of vapor pressure of solvents in pure form and in solutions is a tedious one in Ostwald and Walker's method of determining molecular mass from relative lowering of vapor pressure, the proposed method is easy with simple experimental setup and can be used to determine the molecular weight of substances and polymers soluble in protic solvents.

Abbreviations

EIS	electrochemical impedance spectroscopy
MOF	metal organic framework
R	resistance
C	capacitance
E	voltage
I	current

Author details

Subramaniam Rameshkumar[1]*, Panjaiyan Karthikeyan[2], Iman Danaee[3] and Manogaran Obulichetty[4]

1 Department of Chemistry, Sri Vasavi College, Erode, Tamil Nadu, India

2 Department of Mathematics, Sri Vasavi College, Erode, Tamil Nadu, India

3 Petroleum Engineering, Petroleum University of Technology, Abadan, Iran

4 Department of Chemistry and Applied Chemistry, PSG College of Technology, Coimbatore, India

*Address all correspondence to: srksvc2016@gmail.com

References

[1] Ardavan H. A speed-of-light barrier in classical electrodynamics. Physical Review D. 1984;**29**(2):207

[2] Lai W, Haile SM. Impedance spectroscopy as a tool for chemical and electrochemical analysis of mixed conductors: A case study of ceria. Journal of the American Ceramic Society. 2005; **88**(11):2979-2997

[3] Cole KS, Cole RH. Dispersion and absorption in dielectrics I. Alternating current characteristics. The Journal of Chemical Physics. 1941;**9**(4):341-351

[4] Davidson DW, Cole RH. Dielectric relaxation in glycerol, propylene glycol, and n-propanol. The Journal of Chemical Physics. 1951;**19**(12):1484-1490

[5] Warburg E. Ueber das Verhalten sogenannter unpolarisirbarer Elektroden gegen Wechselstrom. Annalen der Physik. 1899;**303**(3):493-499

[6] Dolin P, Ershler B. The kinetics of discharge and ionization of hydrogen adsorbed at Pt-electrode. Acta Physicochimica URSS. 1940;**13**:747

[7] Randles JEB. Kinetics of rapid electrode reactions. Transactions of the Faraday Society. 1947;**1**:11

[8] Yeager E, Breiter MW. Transactions of the symposium on electrode processes. Journal of The Electrochemical Society. 1962;**109**(1):25C

[9] Macdonald DD. Reflections on the history of electrochemical impedance spectroscopy. In: Paper Presented at 6th International Symposium on Electrochemical Impedance Spectroscopy; 17–21 May 2004; Cocoa Beach, Florida

[10] Mallaiya K, Subramaniam R, Srikandan SS, Gowri S, Rajasekaran N, Selvaraj A. Electrochemical characterization of the protective film formed by the unsymmetrical Schiff's base on the mild steel surface in acid media. Electrochimica Acta. 2011;**56**(11): 3857-3863

[11] Wycisk R, Pintauro PN, Park JW. New developments in proton conducting membranes for fuel cells. Current Opinion in Chemical Engineering. 2014; **4**:71-78

[12] Peighambardoust SJ, Rowshanzamir S, Amjadi M. Review of the proton exchange membranes for fuel cell applications. International Journal of Hydrogen Energy. 2010;**35**(17): 9349-9384

[13] Li Q, Jensen JO, Savinell RF, Bjerrum NJ. High temperature proton exchange membranes based on polybenzimidazoles for fuel cells. Progress in Polymer Science. 2009; **34**(5):449-477

[14] Le HH, Abhijeet S, Ilisch S, Klehm J, Henning S, Beiner M, et al. The role of linked phospholipids in the rubber-filler interaction in carbon nanotube (CNT) filled natural rubber (NR) composites. Polymer. 2014;**55**(18):4738-4747

[15] Phair JW, Badwal SP. Review of proton conductors for hydrogen separation. Ionics. 2006;**12**(2):103-115

[16] Kepert CJ. Advanced functional properties in nanoporous coordination framework materials. Chemical Communications. 2006;**7**:695-700

[17] Rowsell JL, Yaghi OM. Metal–organic frameworks: A new class of

porous materials. Microporous and Mesoporous Materials. 2004;**73**(1-2): 3-14

[18] Maspoch D, Ruiz-Molina D, Veciana J. Old materials with new tricks: Multifunctional open-framework materials. Chemical Society Reviews. 2007;**36**(5):770-818

[19] James SL. Metal-organic frameworks. Chemical Society Reviews. 2003;**32**(5):276-288

[20] Maji TK, Kitagawa S. Chemistry of porous coordination polymers. Pure and Applied Chemistry. 2007;**79**(12): 2155-2177

[21] Brammer L. Developments in inorganic crystal engineering. Chemical Society Reviews. 2004;**33**(8):476-489

[22] Janiak C. Engineering coordination polymers towards applications. Dalton Transactions. 2003;**14**:2781-2804

[23] Zhao D, Timmons DJ, Yuan D, Zhou HC. Tuning the topology and functionality of metal-organic frameworks by ligand design. Accounts of Chemical Research. 2011;**44**(2):123-133

[24] Nagao Y, Kubo T, Nakasuji K, Ikeda R, Kojima T, Kitagawa H. Preparation and proton transport property of *N*, *N′*-diethyldithiooxamidatocopper coordination polymer. Synthetic Metals. 2005;**154**(1-3):89-92

[25] Sadakiyo M, Yamada T, Kitagawa H. Rational designs for highly proton-conductive metal–organic frameworks. Journal of the American Chemical Society. 2009;**131**(29):9906-9907

[26] Yamada T, Sadakiyo M, Kitagawa H. High proton conductivity of one-dimensional ferrous oxalate dihydrate. Journal of the American Chemical Society. 2009;**131**(9):3144-3145

[27] Sahoo SC, Kundu T, Banerjee R. Helical water chain mediated proton conductivity in homochiral metal–organic frameworks with unprecedented zeolitic unh-topology. Journal of the American Chemical Society. 2011; **133**(44):17950-17958

[28] Wong NE, Ramaswamy P, Lee AS, Gelfand BS, Bladek KJ, Taylor JM, et al. Tuning intrinsic and extrinsic proton conduction in metal–organic frameworks by the lanthanide contraction. Journal of the American Chemical Society. 2017;**139**(41): 14676-14683

[29] Ramaswamy P, Wong NE, Shimizu GK. MOFs as proton conductors–challenges and opportunities. Chemical Society Reviews. 2014;**43**(16):5913-5932

[30] Horike S, Umeyama D, Kitagawa S. Ion conductivity and transport by porous coordination polymers and metal–organic frameworks. Accounts of Chemical Research. 2013;**46**(11): 2376-2384

[31] Shimizu GK, Taylor JM, Kim S. Proton conduction with metal-organic frameworks. Science. 2013;**341**(6144): 354-355

[32] Yoon M, Suh K, Natarajan S, Kim K. Proton conduction in metal–organic frameworks and related modularly built porous solids. Angewandte Chemie International Edition. 2013;**52**(10): 2688-2700

[33] Kitagawa H. Transported into fuel cells. Nature Chemistry. 2009;**1**(9): 689-690

[34] Saravanabharathi D, Obulichetty M, Rameshkumar S, Kumaravel M. Humidity based proton conductivity of calcium-L-tartrate tetrahydrate: An environmentally benign coordination polymer as a solid electrolyte. Synthetic Metals. 2014;**196**:76-82

[35] Saravanabharathi D, Obulichetty M, Rameshkumar S, Kumaravel M. Rapid crystallization and proton conductivity of copper (II)-L-tartrate. Synthetic Metals. 2012;**162**(17-18):1519-1523

[36] Saravanabharathi D, Obulichetty M, Rameshkumar S, Kumaravel M. New 2-methyl benzimidazole based zinc carboxylates: Supramolecular structures, biomimetic proton conductivities and luminescent properties. Inorganica Chimica Acta. 2015;**437**:167-176

[37] Singh K, Tiwari RU. Solid state battery of proton conducting sodium thiosulphate pentahydrate. In: Proceedings of the 4th Asian Conference on Solid State lonics. 1994. pp. 403

[38] Chikh L, Delhorbe V, Fichet O. (Semi-)interpenetrating polymer networks as fuel cell membranes. Journal of Membrane Science. 2011;**368**(1-2):1-7

[39] Nasar A, Perveen R. Applications of enzymatic biofuel cells in bioelectronic devices–a review. International Journal of Hydrogen Energy. 2019;**44**(29): 15287-15312

[40] Ye G, Janzen N, Goward GR. Solid-state NMR study of two classic proton conducting polymers: Nafion and sulfonated poly (ether ether ketone)s. Macromolecules. 2006;**39**(9):3283-3290

[41] Coster HG, Chilcott TC, Coster AC. Impedance spectroscopy of interfaces, membranes and ultra structures. Bioelectrochemistry and Bioenergetics. 1996;**40**(2):79-98

Chapter 2

Development of An Impedimetric Nanoplatform for Cu(II) Detection Based on Carbon Dots and Chitosan Nanocomposite

Mosaab Echabaane and Chérif Dridi

Abstract

In this chapter, an impedemitric sensor based on carbon dots (CDs) and chitosan (CS) nanocomposite for Cu (II) detection has been studied. The synthesized carbon dots were characterized by TEM, FTIR, XRD, UV-visible, and PL. The obtained carbon dot/chitosan/GCE structure was characterized by scanning electron microscopy (SEM) and electrochemical impedance spectroscopy (EIS). The behavior, interface processes of GCE/electrolyte, and sensor properties of the studied structures were investigated. In particular, the performance of the proposed impedimetric sensor to Cu ions, such as sensitivity, dynamic range, detection limit, and selectivity, was studied. Subsequently, the proposed sensor was applied for the determination of copper ions in real water samples.

Keywords: carbon dots, chitosan, nanocomposite, copper(II), Impedimetric sensors, introduction

1. Introduction

Essential transition metals such as copper, zinc, iron, and manganese play vital roles in several biological processes [1]. One of the essential transition metal ions, copper is the most abundant metal ion in the human body. It contributes to blood formation, connective tissue development, transcriptional events, and the functioning of several enzymes [2, 3]. However, exposure to an excess level of copper ion can cause gastrointestinal disturbance, neurodegenerative diseases, and liver and kidney damage [4]. The maximum content of Cu^{2+} ions in drinking water is limited to 30 nM by the World Health Organization (WHO) [5].

Hence, it is very important to develop a simple, fast, economically advantageous, selective, and sensitive analytical technique for the determination of copper in aquatic environments to avoid side effects. Notably, numerous techniques have been used to determine the copper ion, including atomic absorption spectroscopy, chromatography spectrofluorimetry, and spectrophotometry [6–9]. These techniques can provide good accuracy and sensitivity, but they have various disadvantages, such as expensive equipment, time-consuming operation, and an inability to do onsite analysis. Thus, their use is restricted. Therefore, to circumvent these limitations, electrochemical sensors have attracted great attention for detecting trace levels of copper ions due to their inherent advantages, such as simple operation, low-cost, rapidity, excellent sensitivity, and selectivity [10]. Notably, the presence of numerous advanced electrode modified nanomaterials improved sensing performance significantly [11–15]. Amongst such nanomaterials, nanocomposites have been employed in electrochemical sensors for environmental analysis [16–21]. Nowadays, analytical measurements are altered by nanocomposites of various sizes, shapes, and compositions. Chitosan as well as carbonaceous nanoparticles have an interest in the construction of electrochemical sensors [22–28]. Chitosan exhibits interesting properties due to its nontoxicity, film-forming ability, high permeability, adsorption ability, and biocompatibility. It has been extensively used as a good biomaterial in designing sensors [29, 30]. As compared to other carbonaceous nanoparticles, carbon dots have considerable importance because of their simple synthetic routes, high surface area, high biocompatibility, and high conductivity [31, 32]. The aim of this chapter is to develop an impedimetric sensor for copper ion detection based on carbon dots and chitosan nanocomposite. The synthesized carbon dots were characterized by transmission electron microscope (TEM), X-ray diffraction (XRD), FTIR spectroscopy, UV–Vis absorption, and photoluminescence spectroscopy. The morphological and electrochemical properties of carbon dot/chitosan/GCE structure were investigated by scanning electronic microscopy (SEM) and electrochemical impedance spectroscopy (EIS). The weather sensors' performance was discussed. The proposed sensor can be used to determine copper ions in real water samples.

2. Experimental details

2.1 Reagents

Copper sulfate ($CuSO_4$, $\geq 99.9\%$), mercury nitrate monohydrate ($Hg(NO_3)_2 \cdot H_2O$, $\geq 99.9\%$), cadmium nitrate tetrahydrate ($Cd(NO_3)_2 \cdot 4H_2O$, $\geq 99\%$), lead nitrate ($Pb(NO_3)_2$, $\geq 99\%$), nickel nitrate hexahydrate ($Ni(NO_3)_2 \cdot 6H_2O$, $\geq 98.5\%$), zinc nitrate hexahydrate ($Zn(NO_3)_2 \cdot 6H_2O$, $\geq 98\%$), sodium chloride ($NaCl$, $\geq 99\%$), calcium chloride ($CaCl_2$, $\geq 96\%$), potassium chloride (KCl, $\geq 99\%$), glucose ($\geq 99.5\%$), sulfuric acid (H_2SO_4, $\geq 99.9\%$), Deionized (DI) water, sodium hydroxide ($NaOH$, $\geq 97\%$), ammonium acetate (CH_3COONH_4, $\geq 98\%$), and chitosan (CS, $\geq 85\%$). All chemicals used in this work were purchased from of Sigma- Aldrich.

2.2 Instrumentations

For further investigation of the synthesized carbon dots (CDs), an X-ray diffractometer (Bruker AXS D8 Advance), Fourier Transform Infrared Spectroscopy (FTIR) (Perkin Elmer, Spectrum Two), UV–vis spectrophotometry (Specord 210 Plus), and a spectrophotometer (Perkin Elmer) were employed.

Scanning Electron Microscopy (SEM) (Jeol JSM-5400 emission) was used to characterize the CS-CDs film.

The electrochemical impedance measurements were carried out using an impedance analyzer (EC-LAB BIOLOGIC) with a sinusoidal excitation signal amplitude of 10 mV and a 0.01 Hz-100 kHz frequency range. The polarization potential used for

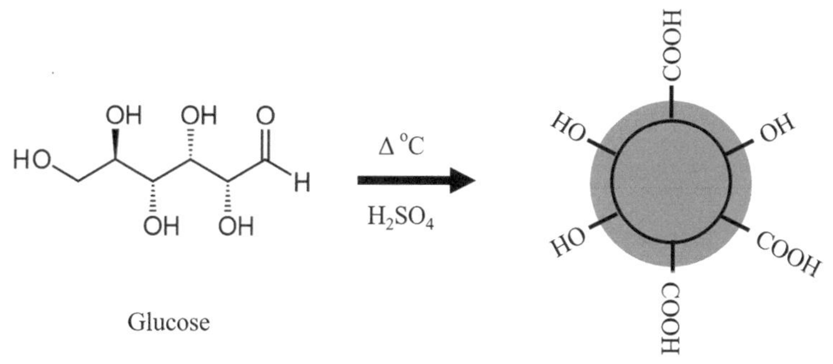

Figure 1.
Schematic representation of CDs formation from glucose.

these measurements was optimized at −600 mv. An aqueous ammonium acetate (solution of 0.1 M, with pH = 7, was used as an electrolyte in the electrochemical cell formed by three electrodes: a working electrode (GCE), a platinum counter electrode, and an Ag/AgCl electrode (SCE).

2.3 Synthesis of CDs

Glucose (3 g) and H_2SO_4 (1 ml) were dissolved in DI water (10 ml). Then, the mixed solution was heated for 30 min at a temperature of 100°C (**Figure 1**). The solution changed from colorless to yellow-brownish, signaling the formation of CDs. Then, this latter was neutralized using NaOH. In order to remove impurities, the prepared solution was centrifuged (14,000 rpm), filtered, and dried for 6 hours at a temperature of 80°C [33].

2.4 Preparation of modified glassy carbon electrode (GCE)

The surface of the GCE electrode was polished with 0.05 μm and 0.3 μm alumina powder. Then, the electrode was washed successively with distilled water, ethanol, and distilled water in an ultrasonic bath, and dried at room temperature. For the preparation of the CDs-CS nanocomposite solution: 0.5 mL of 1.0% CS solution was added to 1.5 mL of CDs solution by ultrasonication. Then, 2 μL of nanocomposite was dropped onto the surface of GCE and dried in the oven at 60°C for 30 min.

3. Characterization of CDs and CS-CDs/GCE

The TEM image of the synthesized CDs is shown in **Figure 2(a)**. It is noticed that the CDs are spherical in shape and their mean size is approximately 3.2 nm. The X-ray diffraction pattern of CDs is shown in **Figure 2(b)**. It exhibited a broad peak at $2\theta = 22.35^{\circ}$, corresponding to the (002) hkl plane of the graphitic structure (JCPDS card no. 26–1076). The FTIR transmittance spectrum of the CDs is presented in **Figure 2(c)**. The peaks at 3319 cm^{-1}, 1727 cm^{-1}, 1642 cm^{-1}, and 1092 cm^{-1} were assigned to O-H, C = O, C = C, and C-O. These indicate that carboxyl and hydroxyl groups may be found on the surface of CDs. The UV–Vis absorption spectrum of CDs exhibited two absorption bands located at 280 nm and 375 nm (**Figure 2(d)**), which are ascribed to

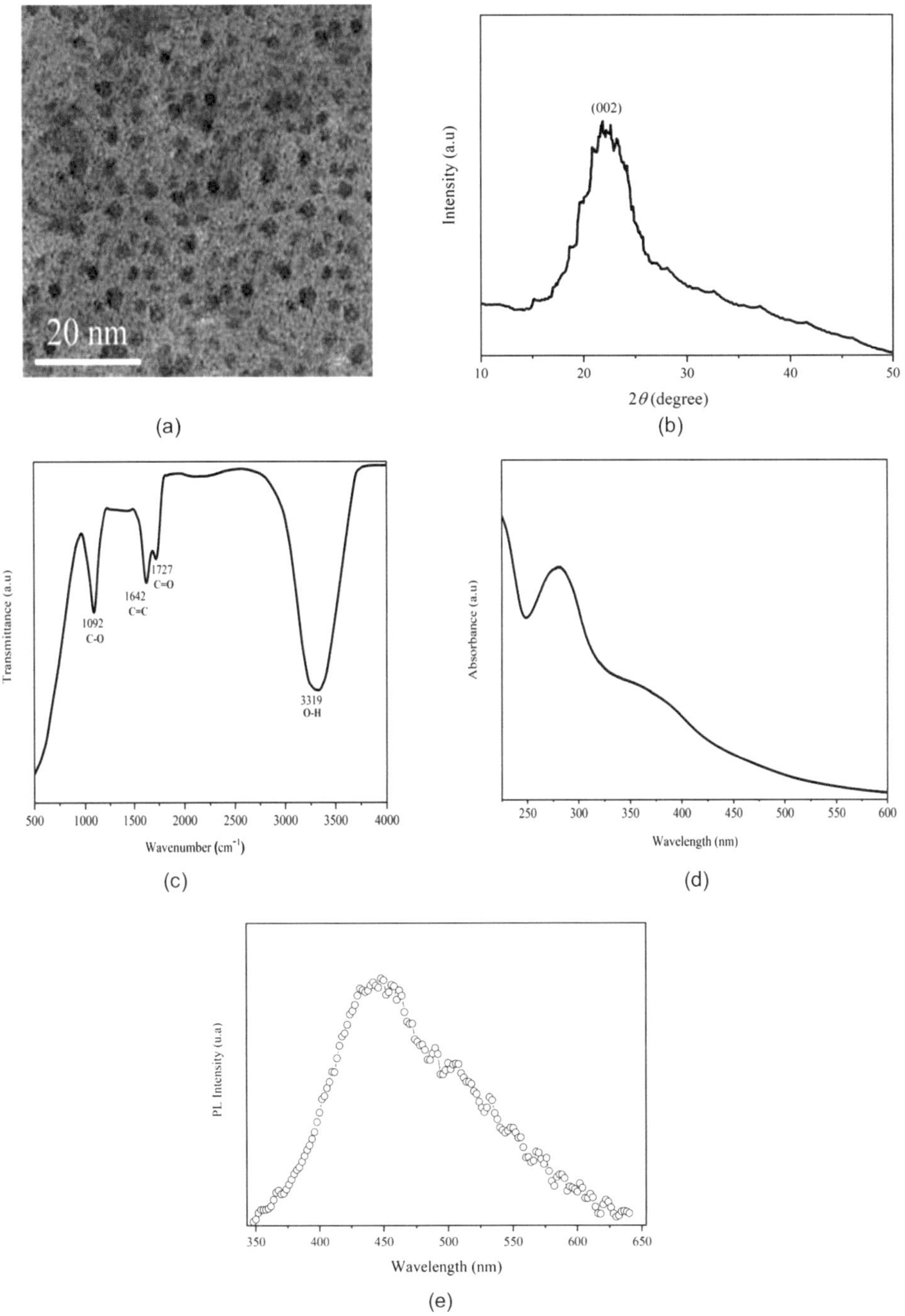

Figure 2.
Characterization of CDs: (a) TEM images, (b) XRD pattern, (c) FTIR spectra, (d) UV–visible spectra, (e) PL spectra.

π-π* and n-π* respectively. **Figure 2(e)** presents the PL spectra of the CDs. The fluorescence emission peak of CDs located at 470 nm is a blue light luminescent spectral region.

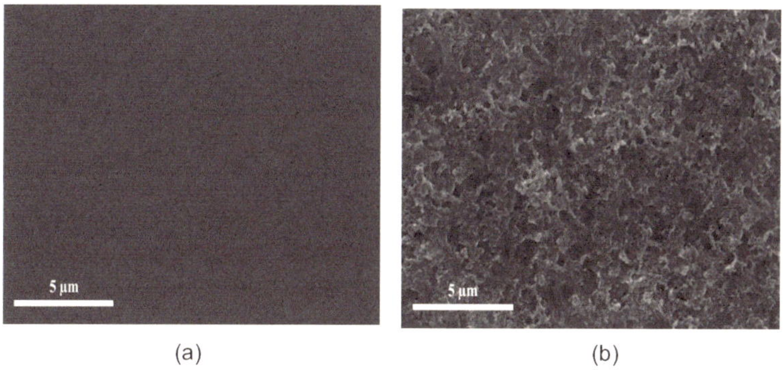

Figure 3.
Images SEM of (a) bare GCE electrode, (b) CS-CDs nanocomposite/GCE electrode.

SEM images of the bare GCE and the CS-CDs/GCE film are shown in **Figure 3**. It exhibits that the bare GCE electrode has a flat and smooth surface (**Figure 3(a)**), whereas the surface of the CS-CDs nanocomposite is rough with a few apparent pores (**Figure 3(b)**). This result shows that the CS-CDs film was deposited on the GCE surface.

4. EIS characterization of the CS-CDs/GCE electrode

Electrochemical impedance spectroscopy is a sensitive tool for investigating the characteristics of the electrode-solution interface. The Nyquist diagrams of the bare GCE electrodes and functionalized CS-CDs film are shown in **Figure 4**. The Nyquist of CS-CDs/GCE electrodes is higher than that of bare CGE electrodes. This is due to the immobilization of the film. Besides, **Figure 4** shows that the Nyquist plot of the CS-CDs/GCE structure is a single semicircle with a large diameter. This might be explained by the fact that it is an overlapping of two semicircles.

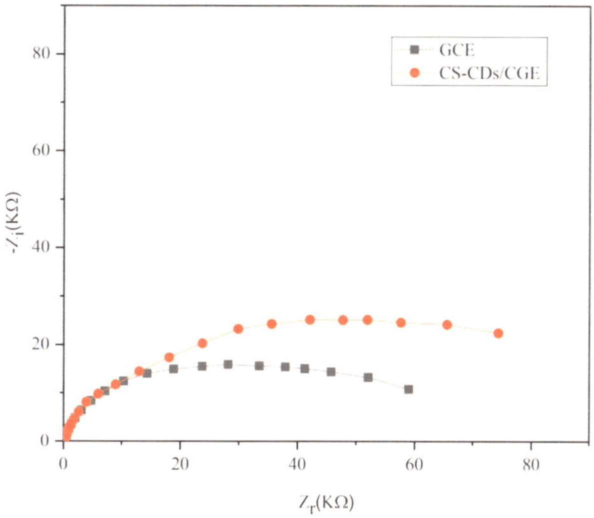

Figure 4.
Impedance spectra of bare GCE, and CS-CDs/GCE structures.

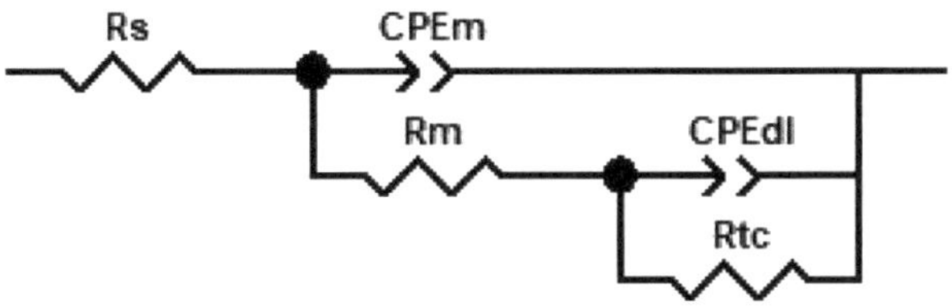

Figure 5.
Equivalent circuit used to fit the impedance spectra.

The EIS spectrum of the CS-CDs/GCE electrode was simulated with the equivalent electrical circuit presented in **Figure 5**. This circuit is an association of three components.

This circuit is an association of three components.

- The first one, in the low-frequency range, is used to represent the GCE/ electrolyte interface, including in a parallel combination the constant phase element CPE_{dl} that substitutes the electrical double layer capacitance (C_{dl}) with the charge transfer resistance R_{ct}.
- The second component at higher frequencies is attributed to the film/electrolyte interface formed by a parallel combination of the resistance R_m and the capacitance CPE_m. Where R_m refers to the resistance of the film and CPE_m donates the constant phase element capacitance, is assigned to the electric capacitor consisting of the metal and the electrolyte, with the film as the dielectric.
- The last component is the series resistance of the electrolyte solution.

The two constant phase elements (CPE) are non-ideal capacitances and can be expressed by (Eq. (1)):

$$Z_{CPE} = 1/_{Q(j\omega)^n} \quad (1)$$

where Q is a constant, j is the imaginary number, ω is the angular frequency, and 0 < n < 1. CPE becomes more capacitive, when the value n tends to be 1.

Finally, the experimental EIS measurements were fitted using the ZView software. The different fits were made with a total error value of $\chi 2 \approx 10^{-4}$.

5. Impedance characterization for the detection of Cu ion

The electrochemical impedance response of copper at the CS-CDs/GCE is characterized using EIS. The Nyquist spectra of the CS-CDs film obtained after the addition of different concentrations of copper ions is shown in **Figure 6**.

It is noted that the Nyquist plot of the CS-CDs film clearly decreases as a function of Cu^{2+} concentration from 10^{-9} M to 10^{-5} M. The response of the CS-CDs modified electrode to copper ions is due to a complexation process. This is induced by coordination between the donor groups of nanocomposites and the copper ions. In fact, the

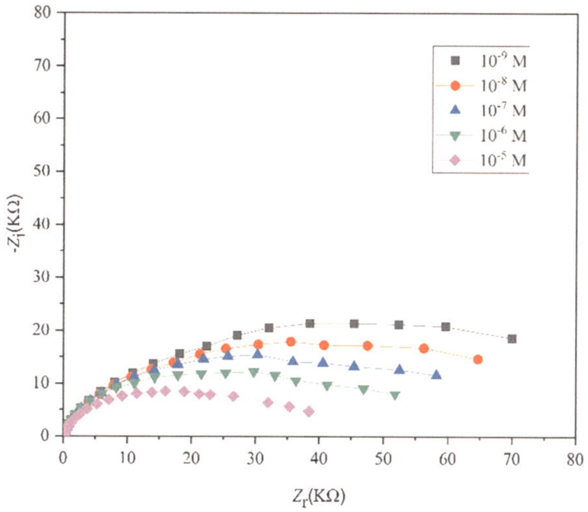

Figure 6.
Impedance spectra for different copper concentrations of CS-CDs/GCE.

$[Cu^{2+}]$ (M)	R_s (kΩ)	CPE_m (μF)	n_1	R_m (kΩ)	CPE_{dl} (μF)	n_2	R_{ct} (kΩ)	χ2 (10^{-4})
10^{-9}	0.11	1.70	0.90	17.40	9.20	0.59	67.24	2.7
10^{-8}	0.10	1.73	0.89	16.24	10.01	0.60	56.56	6.8
10^{-7}	0.11	1.60	0.88	15.01	9.22	0.60	45.05	7.3
10^{-6}	0.12	1.56	0.90	13.40	9.52	0.61	36.16	2.1
10^{-5}	0.13	1.66	0.88	12.50	10.10	0.62	27.72	4.2

Table 1.
Fitting parameters of CS-CDs/GCE versus copper concentrations.

interaction of the copper ions (Lewis acid, acceptor electrons) with the amide and hydroxyl groups of CS (Lewis base, donor electrons) is possible by the acid–base reaction. Furthermore, the negatively charged carboxyl and hydroxyl groups existing on the surface of CDs enable interactions with the positively charged copper ions. Thus, the interaction mechanism is evolved by an electrostatic interaction that occurs between the copper ions and the nanocomposite film.

Table 1 summarizes the circuit parameters of the CS-CDs film as determined by the fit to the previous equivalent circuit.

6. Physical meaning of parameters obtained from the Nyquist plots

Figure 7 shows the plot of the film resistance versus p$[Cu^{2+}]$. The value of film resistance (R_m) decreases when increasing the concentration of copper. This is may be explained by the increase in the ionic conductivity of the CS-CDs film.

Figure 8 represents the variation of the charge transfer resistance (R_{ct}) parameters upon the addition of Cu^{2+}. It was observed that R_{ct} decreased with increasing Cu^{2+} concentration. This can be due to improvement in the charge transfer at the GCE/CS-CDs electrode electrolyte interface.

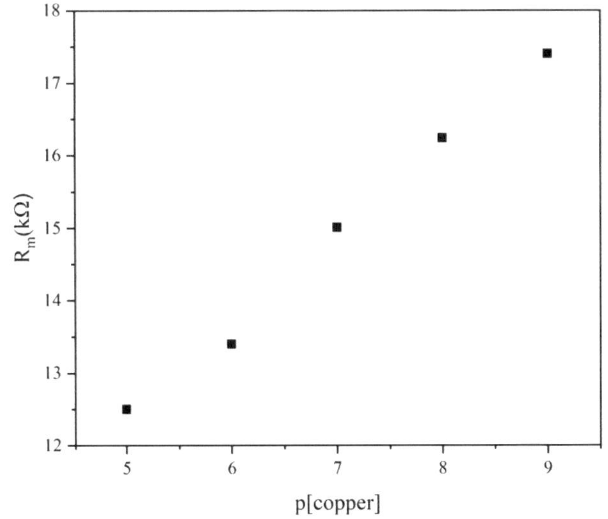

Figure 7.
Variation of the R_m of CS-CDs/GCE.

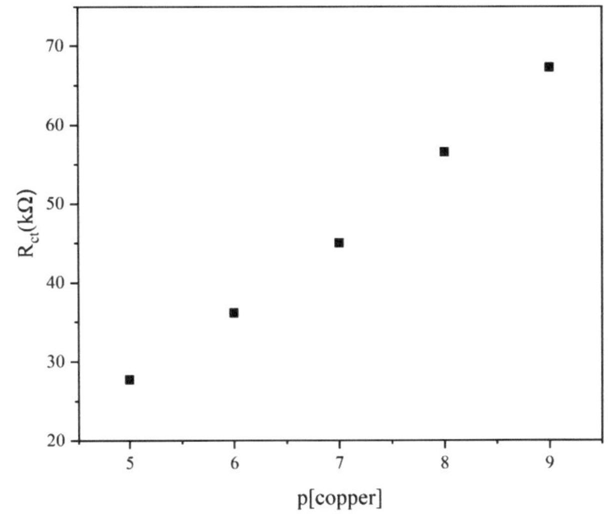

Figure 8.
Variation of the R_{ct} of CS-CDs/GCE.

Electrode/p $[Cu^{2+}]$	R_m (KΩ)	R_{ct} (KΩ)
CS-CDs/GCE	1.3	9.9

Table 2.
R_m and R_{ct} slope values of CS-CDs/GCE.

Table 2 presents the R_m and R_{ct} slope values. It is noted that the R_{ct} slope value is higher than the R_m slope value. This can indicate that the recognition process was at the electrolyte/GCE interface. Therefore, the R_{ct} is provided as the primary factor determining the sensors' response.

7. Analytical performance of impedimetric sensors

Figure 9 shows the variation of -log(R_{tc}/R_{tc0}) as a function of copper concentration, where R_{tc0} is the charge transfer resistance value without copper ions in the solution. Here, a linear behavior with a large detection range between 10^{-9} and 10^{-5} M was observed. The equation of linear regression is -log (R_{tc}/R_{tc0}) = 0.1 × log [copper] + 0.9, where the slope is 0.1 of the sensor sensitivity. The detection limit of our copper sensor was ≈ 5×10^{-10} M, according to the formula LOD = (3 × standard deviation)/slope.

8. Specificity

The specificity of the GCE/CS-CDs sensor has been studied towards some interfering cations: Hg^{2+}, Cd^{2+}, Pb^{2+}, Ni^{2+}, Zn^{2+}, Na^{+}, Ca^{+}, and K^{+}. **Figure 10** shows the relative signal change $\left(\left[\frac{R_{tc}-R_{tc0}}{R_{tc0}}\right]\right)$ as a function of the interfering cations. The results noted that GCE/CS-CDs displayed good selectivity for Cu^{2+} in the presence of other ions.

9. Determination of copper ion in water samples

To evaluate the practical feasibility of the GCE/CS-CDs sensor, it is applied for the detection of copper in water samples (tap water). 1 mL of sample solution (tap water) is added to 1 mL of electrolyte (acetate ammonium at pH 7). The impaled copper amounts are of 10^{-6} M and 10^{-8} M. The recovery percentage is calculated as $\frac{Found\ (M)}{Added\ (M)} \times 100$. It can be seen in **Table 3** that the recoveries are less than 5%. Therefore, the proposed GCE/CS-CDs impedimetric sensor is efficient for the detection of Cu^{2+} ions in real water samples.

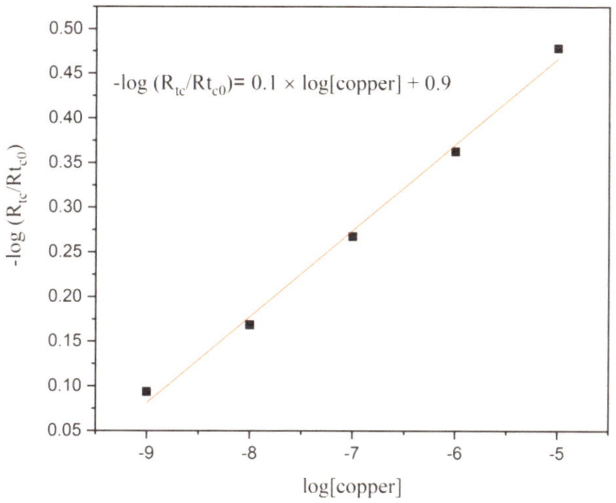

Figure 9.
Calibration curve of -log (R_{tc}/R_{tco}) versus log ([copper]).

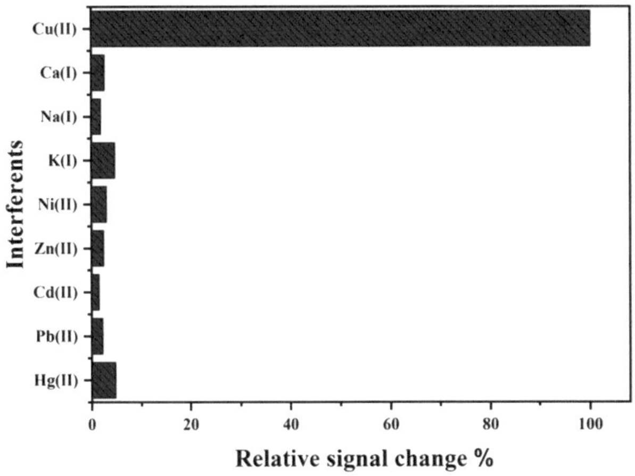

Figure 10.
Influence of interferences (10^{-4} M) on the response copper (10^{-6} M).

Sample	Added (M)	Found (M)	Recovery (%)	RSD (%)
Tap water	10^{-8}	9.70×10^{-9}	97.0	4.12
	10^{-6}	1.02×10^{-6}	102. 0	1.47

Table 3.
Determination of copper ion in tap water samples (n = 3).

10. Conclusion

In this chapter, we fabricate a highly sensitive, selective, impedimetric sensor for the determination of copper-based on carbon dots/chitosan nanocomposite modified glassy carbon electrode. A simple, fast, low-cost method to prepare carbon dots using glucose by heating has been demonstrated. The electrochemical performance of the copper sensor was examined by electrochemical impedance spectroscopy. Based on the excellent properties of CS-CDs/GCE, the impedimetric sensors had high sensitivity, a low limit of detection, and good selectivity. The newly developed impedimetric sensor for the determination of copper ions was applied in water for real samples with satisfactory results. Thus, this indicates that these sensors offer promising potential for water monitoring.

Author details

Mosaab Echabaane* and Chérif Dridi
NANOMISENE Lab, LR16CRMN01, Centre for Research on Microelectronics and Nanotechnology CRMN of Technopark of Sousse, Sousse, Tunisia

*Address all correspondence to: mosaab.echabaane@gmail.com

References

[1] Cowan JA. Handbook of Inorganic Biochemistry: An Introduction. 2nd ed. New York: Wiley; 1997. 456 p. DOI: 978-0-471-18895-7

[2] Nasiri MM, Taher MA, Fazelirad H. Preparation and application of a simple electrochemical sensor for the determination of copper in some real and standard samples. Ionics. 2016;**22**: 289-296. DOI: 10.1007/s11581-015-1533-9

[3] Rouis A, Echabaane M, Sakly N, Bonnamour I, Ouada HB. Characterization of a sensitive and selective copper optode based on β-ketoimine modified calix[4]arene derivative. Materials Science and Engineering: C. 2015;**46**:125-131. DOI: 10.1016/j.msec.2014.10.026

[4] Ansari R, Delavar AF, Aliakbar A, Mohammad-khah A. Solid-state Cu (II) ion-selective electrode based on polyaniline-conducting polymer film doped with copper carmoisine dye complex. Journal of Solid State Electrochemistry. 2012;**16**: 869-875. DOI: 10.1007/s10008-011-1436-2

[5] Wu T, Xu T, Ma Z. Sensitive electrochemical detection of copper ions based on the copper (II) ion assisted etching of Au@ Ag nanoparticles. Analyst. 2015;**140**:8041-8047. DOI: 10.1039/C5AN01888A

[6] Soylak M, Narin I, Dogan M. Trace enrichment and atomic absorption spectrometric determination of lead, copper, cadmium and nickel in drinking water samples by use of an activated carbon column. Analytical Letters. 1997; **30**:2801-2810. DOI: 10.1080/00032719708001823

[7] Ali A, Shen H, Yin X. Simultaneous determination of trace amounts of nickel, copper and mercury by liquid chromatography coupled with flow-injection on-line derivatization and preconcentration. Analytica Chimica Acta. 1998;**369**:215-223. DOI: 10.1016/S0003-2670(98)00252-9

[8] Amjadi M, Fakhri ZA. Gold nanostar-enhanced chemiluminescence probe for highly sensitive detection of Cu (II) ions. Sensors and Actuators B: Chemical. 2018;**257**:629-634. DOI: 10.1016/j.snb.2017.11.013

[9] Agnihotri NK, Singh VK, Singh HB. Derivative spectrophotometric determination of copper (II) in non-ionic micellar medium. Talanta. 1997;**45**: 331-341. DOI: 10.1016/S0039-9140(97)00137-9

[10] Kaur I, Sharma M, Kaur S, Kaur A. Chemical sensors based on self-assembled chelating dithiol on gold electrode for trace level detection of copper (II) ions. Sensors and Actuators B: Chemical. 2020;**312**: 127935-127946. DOI: 10.1016/j.snb.2020.127935

[11] Mahmoud A, Echabaane M, Omri K, ElMir L, Chaabane RB. Development of an impedimetric non enzymatic sensor based on ZnO and Cu doped ZnO nanoparticles for the detection of glucose. Journal of Alloys and Compounds. 2019;**786**:960-968. DOI: 10.1016/j.jallcom.2019.02.060

[12] Somashekharappa KK, Rajendrachari S. Handbook of Carbon Nanomaterial-Based Sensors. 1st ed. Elsevier. 2022. p. 3. DOI: 10.1016/B978-0-323-91174-0.00001-9

[13] Rajendrachari S, Ramakrishna D. Handbook of Functionalized Nanomaterial-Based Electrochemical Sensors. 1st ed. Elsevier; 2022. p. 3. DOI: 10.1016/B978-0-12-823788-5.00010-7

[14] Shashanka R, Jayaprakash GK, Prakashaiah BG, Kumar M, Swamy BEK. Electrocatalytic determination of ascorbic acid using a green synthesised magnetite nano-flake modified carbon paste electrode by cyclic voltammetric method. Materials Research Innovations. 2022;**26**:229-239. DOI: 10.1080/14328917.2021.1945795

[15] Shashanka R, Swamy BEK. Simultaneous electro-generation and electro-deposition of copper oxide nanoparticles on glassy carbon electrode and its sensor application. SN Applied Sciences. 2020;**2**:956. DOI: 10.1007/s42452-020-2785-1

[16] Jebril S, Aguilera LC, Santander JMP, Dridi C. A novel electrochemical sensor modified with green gold sononanoparticles and carbon black nanocomposite for bisphenol a detection. Materials Science and Engineering B. 2021;**264**:114951-114961. DOI: 10.1016/j.mseb.2020.114951

[17] Raghu GK, Sampath S, Pandurangappa M. Chemically functionalized glassy carbon spheres: A new covalent bulk modified composite electrode for the simultaneous determination of lead and cadmium. Journal of Solid State Electrochemistry. 2012;**16**:1953-1963. DOI: 10.1007/s10008-011-1595-1

[18] Yin T, Pan D, Qin W. A solid-contact Pb^{2+} –selective polymeric membrane electrode with nafion-doped poly (pyrrole) as ionto-electron transducer. Journal of Solid State Electrochemistry. 2012;**16**:499-504. DOI: 10.1007/s10008-011-1358-z

[19] Lisak G, Bobacka J, Lewenstam A. Recovery of nanomolar detection limit of solid-contact lead (II)-selective electrodes by electrode conditioning. Journal of Solid State Electrochemistry. 2012;**16**:2983-2299. DOI: 10.1007/s10008-012-1725-4

[20] Intarakamhang S, Schuhmann W, Schulte A. Robotic heavy metal anodic stripping voltammetry: Ease and efficacy for trace lead and cadmium electroanalysis. Journal of Solid State Electrochemistry. 2013;**17**:1535-1542. DOI: 10.1007/s10008-013-2018-2

[21] Özcan A, Hamid F, Özcan AA. Synthesizing of a nanocomposite based on the formation of silver nanoparticles on fumed silica to develop an electrochemical sensor for carbendazim detection. Talanta. 2021;**222**:121591-112160. DOI: 10.1016/j.talanta.2020.121591

[22] Rizwan M, Elma S, Lim SA, Ahmed MU. AuNPs/CNOs/SWCNTs/chitosan-nanocomposite modified electrochemical sensor for the label-free detection of carcinoembryonic antigen. Biosensors & Bioelectronics. 2018;**107**:211217. DOI: 10.1016/j.bios.2018.02.037

[23] Jiang H, Du C, Zou Z, Li X, Akins DL, Yang H. A biosensing platform based on horseradish peroxidase immobilized onto chitosan-wrapped single-walled carbon nanotubes. Journal of Solid State Electrochemistry. 2009;**13**:791-798. DOI: 10.1007/s10008-008-0612-5

[24] Li Y, Li Y, Yang Y. A new amperometric H_2O_2 biosensor based on nanocomposite films of

chitosan6MWNTs, hemoglobin, and silver nanoparticles. Journal of Solid State Electrochemistry. 2012;**16**: 1133-1140. DOI: 10.1007/s10008-011-1503-8

[25] Wang L, Zhu H, Hou H, Zhang Z, Xiao X, Song Y. A novel hydrogen peroxide sensor based on Ag nanoparticles electrodeposited on chitosan-graphene oxide/cysteamine-modified gold electrode. Journal of Solid State Electrochemistry. 2012;**16**: 693-1700. DOI: 10.1007/s10008-011-1503-8

[26] Huang J, Xie Z, Huang Y, Xie L, Luo S, Fan Q, et al. Electrochemical immunosensor with Cu (I)/Cu (II)-chitosan-graphene nanocomposite-based signal amplification for the detection of Newcastle disease virus. Scientific Reports. 2020;**10**:1-12. DOI: 10.1038/s41598-020-70877-3

[27] Comino CS, Rassas I, Minot S, Bessueille F, Arab M, Chevallier V, et al. Voltammetric sensor based on molecularly imprinted chitosan-carbon nanotubes decorated with gold nanoparticles nanocomposite deposited on boron-doped diamond electrodes for catechol detection. Materials. 2020;**13**: 688-699. DOI: 10.3390/ma13030688

[28] Jalali F, Hassanvand Z, Barati A. Electrochemical sensor based on a nanocomposite of carbon dots, Hexadecyltrimethylammonium bromide and chitosan for Mesalazine determination. Journal of Analytical Chemistry. 2020;**75**:544-552. DOI: 10.1134/S1061934820040061

[29] Nazari F, Ghoreishi SM, Khoobi A. Bio-based Fe3O4/chitosan nanocomposite sensor for response surface methodology and sensitive determination of gallic acid. International Journal of Biological Macromolecules. 2020;**160**:456-469. DOI: 10.1016/j.ijbiomac.2020.05.205

[30] Sheng Q, Liu R, Zheng J, Lin W, Li Y. Direct electrochemistry of glucose oxidase immobilized on porous carbon nanofiber/room temperature ionic liquid/chitosan composite film and its biosensing application. Journal of Solid State Electrochemistry. 2012;**6**:739-745. DOI: 10.1007/s10008-011-1420-x

[31] Wang X, Feng Y, Dong P. Huang J a mini review on carbon quantum dots: Preparation, properties and electrocatalytic application. Frontiers in Chemistry. 2019;**7**:671-680. DOI: 10.3389/fchem.2019.00671

[32] Campuzano S, Sedeño PY, Pingarrón JM. Carbon dots and graphene quantum dots in electrochemical biosensing. Nanomaterials. 2019;**9**: 634-652. DOI: 10.3390/nano9040634

[33] Echabaane M, Hfaiedh S, Smiri B, Saidi F, Dridi C. Development of an impedimetric sensor based on carbon dots and chitosan nanocomposite modified electrode for Cu(II) detection in water. Journal of Solid State Electrochemistry. 2021;**25**:1797-1806. DOI: 10.1007/s10008-021-04949-3

Section 2

Nanoparticle Enhanced Voltammetric Applications

Chapter 3

An Overview of the Synergy of Electrochemistry and Nanotechnology for Advancements in Sensing Applications

Rajni Bais

Abstract

Electrochemical sensors have been widely employed in diverse domains of electrochemical analysis, biosensing, drug administration, healthcare, agriculture, and so on because of their special potential features that are closely related to their high selectivity, sensitivity and cycling stability. Various electrochemical techniques employed to transduct biological or chemical signal to electrical signal are voltammetry, conductometry, potentiometry and amperometry. Due to the high demand of global market and human interest in having a device to check the concentration of species in different samples that is simple and fast, researchers have been engaged in a fierce competition to design and build new sensors and biosensors in recent years. The performance of the sensors can be considerably improved by modifying the electrode surfaces using diverse nanomaterials. Further, electrochemical biosensors are promising diagnostic tools that can find biomarkers in bodily fluids including sweat, urine, blood or excrement. Nanoparticles have found propitious role in biosensors, because they aid in functions like immobilisation of molecules, catalysis in electrosynthesis, facilitation of electron transfer between electrodes and biomolecules and labelling of biomolecules. The advance in the research amalgamating electrochemistry and nanotechnology for electro (bio) sensing applications is the beginning of a promising future for mankind and global market.

Keywords: nanomaterials/ nanoparticles (NPs), electrochemical, sensor, biosensor

1. Introduction

The information carried by an electrochemical redox reaction (the reaction between electrode and analyte) is converted into an applicable quantitative or qualitative signal by electrochemical sensors. The sensors can generate electronic outputs in the form of digital signals that can be analysed further [1]. Depending on the mechanism involved in recognition, the sensors can be chemical or biological (biosensor). Electrochemistry is found to play a crucial role in medical technology, forensics, food, environmental sciences, defence settings, agriculture, among others [2, 3].

Electrochemical sensors and detectors are very appealing for applications, viz. on-site monitoring of environmental contaminants, addressing varied environmental needs, health monitoring, testing of adulterants in food samples, forensic drug analysis, and many more. Many of the specifications for on-site environmental scanning are met by sensing devices. They are intrinsically sensitive and selective to electroactive species and are fast, reliable, portable, compact and economical [4]. A transducer and a chemical recognition system are the two primary components of an electrochemical sensor, and collectively these two components constitute a sensing electrode. Other electrodes may also be used in electrical measurements like reference and counter electrodes (CEs) [5]. In electrochemical sensors, a potential difference is applied in between working and counter electrodes, and on the basis of redox reactions caused by the analyte at the electrode's interface, the resulting current response is measured. Voltammetry, conductometry, potentiometry, or amperometry can be used to study the transduction of biological or chemical signal to electrical signal. Since their invention, the electrochemical sensors are being studied extensively for their applications in efficient biosensors, immunosensors, special electrode design automated systems, and microelectrodes, etc. [6]. With the recent advancements in the designing and synthesis of nanomaterials along with the evaluation of intrinsic properties of nanoparticles (NPs) based on carbon [7] and many other types of base materials, the engine of nano-electrochemistry has begun to gain speed in the field of electrosensing.

2. Electrochemical signal transduction techniques

A transduction process involves efficient capture of biological or chemical recognition signals and their conversion into electrical, gravimetric, optical, electrochemical or acoustic signals. There are four main types of electrochemical signal transduction techniques that can be used in analysis, depending on the reaction under investigation:

i. Voltammetry

ii. Potentiometry

iii. Amperometry

iv. Conductometry

Each of these techniques is described in the following section in brief.

2.1 Voltammetry

Voltammetry is an electroanalytical method that obtains information about an analyte by modifying a potential and then measuring the resultant current. Here, the potential sweep is applied between working electrode (WE) and counter electrode (CE) with respect to reference electrode (RE), and the current so produced is measured as an analytical signal. Because the potential can be varied in many ways, many forms of voltammetry are available, such as direct current (DC) polarography, differential pulse polarography (DPP), differential staircase voltammetry, cyclic voltammetry, linear sweep voltammetry, normal pulse, and reverse pulse voltammetry [8–10]. Cyclic voltammetry is one of the most widely used electroanalytical methods.

2.2 Potentiometry

In potentiometry, the analytical signal is the open-circuit voltage between working electrode and reference electrode. Depending on the concentration of the analyte, this signal can increase or decrease [11]. Here, Nernst equation governs the relationship between concentration and potential. Useful information about ion activity in redox reaction is provided by potentiometry [12]. Ion-selective electrodes (ISEs) are often employed to achieve low detection limits for potentiometric sensors. Also the potentiometric sensors are ideal for measuring low concentrations of analyte in small sample volumes, because they do not chemically influence the sample [13].

2.3 Amperometry

It is similar to voltammetry with the difference that here a constant or stepped potential is employed to measure the current as electric signal, whereas in voltammetry, controlled variation of voltage is done to measure current. In fact, some scientists classify voltammetry as a type of amperometry only, because both these techniques employ measurement of current upon variation of potential. Here, a continuous measurement of current is done, and this current is produced by oxidation or reduction of an analyte in a biochemical reaction [9, 14]. Here, the peak value of the measured current (over a linear range of potential) is a direct indication of the bulk concentration of the electroactive species [15, 16]. Glucose biosensor is one of the most widely employed amperometric sensors. It is often claimed by many scientific practitioners that amperometric sensors are superior to potentiometric sensors when it comes to sensitivity [17].

2.4 Conductometry

Conductometry sensors are used to measure concentrations of electrolyte in aqueous solutions. The recorded electrical resistance of the solution is used to calculate the molar concentration of an identified electrolyte that causes solution conductivity [18]. Conductivity can be measured directly with a conductivity metre or indirectly with conductometric titration. Electrolyte conductometric analysis has long been used. Conductometric methods were used to analyse mineral waters and salt solutions by Henry Cavendish and Andreas Baumgartner [19].

3. Types of sensors

Sensors can be classified into two broad categories depending on their application or mode of transduction, as follows:

3.1 Types of sensors on the basis of application

i. **Chemical Sensors:** A chemical sensor converts an analyte's physical and/ or chemical properties into a measurable signal [20]. The intensity of the measurable signal is normally proportional to concentration of the analyte [21]. The chemical sensors can further be classified as follows:

a. **pH Sensors:** These sensors are sensitive to even minute changes in pH of a solution and are used to measure acidity and alkalinity in aqueous and other solutions. When used properly, these devices can guarantee the safety and quality of a product as well as the processes that take place in a wastewater treatment or different manufacturing units in industries [22].

b. **Gas Sensors:** These sensors are employed to detect the type and amount or concentration of a gas in a specific environment. Any change in the composition of gas can be linked with electric signals through these sensors [23].

c. **Alcohol Sensors:** Technically known as a MQ3 sensor, an alcohol sensor is a non-contact breath sensor that is mostly used to detect ethanol in the air. When a drunk person breathes near an alcohol sensor, the sensor detects the ethanol in his breath and provides an output based on the concentration of alcohol [24].

d. **Ion-selective Sensors:** Ion-selective electrodes are a type of potentiometric devices. Many electrode systems exhibit a Nernstian relationship between the activity of a redox species in solution and measured electrode potential. The affinity of surface of the membrane for a typical redox species as well as minimum ion conductivity over the membrane are important requirements for the advancement of ion-selective electrodes [25].

e. **Humidity Sensors:** A humidity sensor is a device that detects, measures, and reports the relative humidity (RH) of air or the amount of water vapour present in a gas mixture (air) or pure gas. Humidity sensing is associated with the adsorption and desorption of water. Industrial and agricultural products are both monitored by humidity sensors. Incubators, sterilisers, and pharmaceutical processing plants all use humidity sensors [26].

ii. **Biosensors:** Biosensor is a powerful, efficient, and innovative analytical device that incorporates a biological sensing element and has a wide range of applications, like, environmental monitoring, diagnosis, biomedicine, food safety and processing, drug discovery, defence, and security [27–29]. In a biosensor, an electron exchange takes place in between layer of bioreceptor and an electrode, and this exchange of electrons (because of reduction or oxidation of the analyte) can be transduced into a readable signal [30].

3.2 Types of sensors on the basis of mode of transduction

i. **Amperometric Sensors:** The working of amperometric sensor is based on measuring current generated because of an electrochemical reaction at the electrode surface when a constant working potential is applied with respect to the reference electrode [31]. The amperometric set-up may consist of either two-electrode (WE and RE) or three-electrode (WE, RE, and CE) assembly.

ii. **Potentiometric Sensors:** These sensors are used to measure difference in potential applied between two electrodes when there is no current flowing between them. This measured potential is then used to detect or quantify the analyte of interest in the solution. In a general potentiometric sensor, the signal obeys Nernst Eq. [32].

iii. **Optical Sensors:** These are light-based sensors, which work on the principle of studying the change in wavelength of a particular light when the photosensitive analyte in the medium interacts with the recognition element [33]. Position-sensitive diode, phototransistor, photoresistor, photodiode, and diode array are some commonly used light sensors [34].

iv. **Piezoelectric Sensors:** In these sensors, environmental or mechanical energy changes are converted into electric signals or vice versa. The environmental or mechanical energy changes may include alteration in strain, vibration, pressure, force, etc. [35]. Piezoelectric sensors can be of two types, i.e. passive or active.

v. **Thermal or Calorimetric Sensors:** In these sensors, the concentration of analyte is correlated to amount of heat produced during a molecular recognition reaction. The working principle of calorimetric sensors is to detect the presence or measure the concentration of an analyte by studying the changes in enthalpy produced as a result of any physisorption process or chemical reaction in the medium under observation [36]. Calorimetric sensors are also known as chemoresistors, and the two types of chemoresistors are low-temperature chemoresistors and high-temperature chemoresistors.

4. Types of nanoparticles used in electrosensing

A nanoparticle has diameter approximately one lakh times smaller than that of a human hair strand. Nanomaterials, like in many other technological or scientific segments, have demonstrated their applicability and worthiness for electrochemical/biosensing applications. The astute use of such nanomaterials has resulted in clearly improved performance, with enhanced specificity, sensitivity, and extremely lowered limits of detection. Further, the high specific surface area of all nano-objects allows for the immobilisation of an increased number of bioreceptor units (in biosensors) [37]. The direct functionalisation of nanomaterials (during their synthesis) or coating of nanomaterials with specific functional polymers results in nanomaterials equipped with desired functions without altering their specific properties [38]. The commonly used nanoparticles in electrosensing applications are of following types:

i. **Gold nanoparticles (AuNPs)**

ii. **Quantum dots (QDs)**

iii. **Magnetic nanoparticles (MNPs)**

iv. **Carbon nanotubes (CNTs)**

A brief description of these nanoparticles is given as follows.

4.1 Gold nanoparticles

a. *Properties:* The characteristics of AuNPs, such as their optical and electronic properties, their biocompatibility, and their comparatively simple production and modification, make them of high priority and utility among the class of noble metal nanoparticles [38, 39]. The electrons of AuNPs when irradiated with light of a specific wavelength oscillate in the conduction band (resonant surface plasmons), and this optical behaviour of AuNPs makes them of immense applicability in the field of optical sensing. The use of surface plasmons resonance (SPR) transduction for sensing applications of gold nanoparticles is based on studying the changes in environment of gold films with reference to dielectric constant of propagating surface plasmons. Here, variation in intensity, angle or phase of the reflected light is used to monitor the analyte [40]. Phyto-AuNPs that are phytosynthesised with the aid of green chemistry approach are catalytically and biologically active, biocompatible, and stable. But despite their potent applications, these have not been explored sufficiently [41].

b. *Recent Research Trends:* Owing to the aforementioned properties, the use of AuNPs for the development of novel sensing techniques holds significant promises when it comes to research related to: drug determination [42–44]; anticancer and other biomedical applications [45, 46]; immuno-PCR (Polymerase Chain Reaction) assay and ELISA (Enzyme-Linked Immunosorbent Assay) [47, 48], and so on.

4.2 Quantum dots

a. *Properties:* QDs are nanoscale crystals created by scientists which are capable of electron transport. These semiconducting QDs show the property of emitting a variety of colours when UV light strikes them. These semiconductor nanoparticles have found varied applications in fluorescent biological labels, electrochemical sensing, solar cells and composites. The optical properties of a particle can be altered by controlling their size, and so, these particles can also be tuned or regulated in a way that they can absorb or emit light (colours) of specific wavelength [49]. Further, the shape (hollow/ solid), composition and structure are other crucial parameters of QDs that can affect their various properties. Another important application of QDs is as single-electron transistors. But studies have shown that these semiconductor nanoparticles may exert harmful effects on living systems in some cases [50].

b. *Recent Research Trends:* When carbon QDs are doped with heteroatoms [51], such as nitrogen [52], both sulphur and nitrogen [53], zinc [54], lanthanum [55] and $CuInS_2$ [56], then their optical/photoluminescent properties are found to improve prominently for sensing applications.

4.3 Magnetic nanoparticles

a. *Properties:* MNPs can be assimilated into transducer materials or scattered in the sample prior to being attracted to the active recognition surface of the sensor by an external magnetic field [57]. Because of the reduced number of magnetic domains in nanosized MNPs, these exhibit magnetic behaviours different than the bulk material, resulting in superparamagnetic behaviour. This means that in a very short time, magnetisation can rapidly and randomly flip directions, and when an external magnetic field is absent, the magnetisation denotes to be average zero. This is a temperature-dependent phenomenon which disappears when the magnetic moments are aligned by the application of an external magnetic field [58]. MNPs offer a highly sensitive technique of transduction in biosensors, optical sensors and electrochemical sensors [57, 58].

b. *Recent Research Trends:* In the recent times, the utility of MNPs has increased multifold. Some advanced sensing applications of MNPs are as follows: diagnosis of anticancer drug 6-mercaptopurine using cerium-based MNP as a fast-response and efficient fluorescence quenching sensor [59], Ihlamur leaves [60] and saffron flowers [61]-based biosynthesis of magnetic nanoparticles for antibacterial applications, use of silver MNPs to enhance the surface plasmon resonance signal for determination of leukocyte cell-derived chemotaxin-2 which is an important biomarker for the diagnosis of liver fibrosis [62], green synthesis of Fe_3O_4 nanoflakes and its utility as an electrocatalyst for the voltammetric determination of ascorbic acid [63], diversified biomedical applications of multifunctional MNPs [64], and many more.

4.4 Carbon nanotubes

a. *Properties:* Owing to the distinctive electronic properties (high conductivity and large surface-to-volume ratios) of CNTs, in recent years these have been widely investigated for their utility as electrosensing devices. Non-functionalised CNTs are not very selective for many chemicals, but functionalisation with different chemical moieties can be done to surpass this drawback. For their electrosensing applications, the CNTs can be functionalised with specific metals or functional groups [65]. The adsorption and binding of specific molecules on the surface of carbon nanotubes result in change in potential– current curves, and this forms the working principle of these devices [66]. The properties of CNTs as sensors are mainly dependent on their shape. The diameter of CNTs is of one to several nanometres, and their length may be of several microns. Well-ordered hexagonal carbon cycles or graphite rolls constitute their surface [67]. One to multilayered tubulenes (with closed or open terminations) may be formed, depending on the conditions of synthesis of these nanotubes.

b. *Recent Research Trends:* The characteristic properties of CNTs allow nanotechnology to provide promising platforms for sensing applications. Some of the emerging researches done in this field are as follows: designing of terahertz metasurface-based single-walled CNT [68], synthesis of sensing piezoresistive materials with carbon nanotubes [69], diagnosis of pancreatic and liver cancer employing CNTs [70], etc.

5. Role or functions of nanoparticles in (bio) sensing techniques

The various roles of NPs in (bio) sensing techniques are as follows.

i. **Immobilisation of molecules:** Chemical functional groups or biomolecules (antibody, DNA, enzymes, and cells) that are immobilised on transducer surface during the construction of a (bio) sensor decide for its stability, sensitivity, selectivity and reproducibility [71]. Hence, a number of methods are employed for immobilisation of desired (bio) molecules onto solid support materials. Some of these methods are adsorption, covalent bonding, entrapment, copolymerisation or crosslinking and encapsulation [72].

ii. **Catalysis in electrosynthesis/electrochemical reactions:** The use of solid nanoparticles as heterogeneous catalysts during electrosynthesis reactions has increased in recent times. Under green chemistry conditions, nanocatalysts can be employed efficiently to promote electrochemical reactions [73]. The incorporation of catalytic nanoparticles into electrochemical (bio) sensors can reduce overpotentials in many analytically crucial electrochemical reactions. Also the reversibility of some otherwise irreversible redox reactions (at traditional unmodified electrodes) can be enabled by nanocatalysts [74]. Production of hydrogen through splitting of water by renewable electricity [75], reduction of carbon dioxide into hydrocarbon products [73] and conversion of biomass products to high-value chemicals [73] are some of the commercially or energetically important electrochemical reactions that use nanocatalysts. Nanoparticles of metals like platinum, gold, silver, palladium, copper, nickel and iridium and also some oxide NPs have been used as catalysts in electrochemical reactions [74].

iii. **Enhancement of electron transfer between electrode surfaces and proteins:**

iv. The conductivity properties of metallic nanoparticles make them suitable for facilitating electron transfer between electrodes and active centres of proteins/enzymes, thereby ascertaining their role as 'electrical wires' or 'mediators' [74].

v. **Labelling of biomolecules:** The use of NPs for labelling biomolecules aids in retaining their interaction with cellular/molecular counterparts and also in maintaining their bioactivity. Stripping voltammetric technique is commonly used to measure the trace amounts of dissolved nanoparticle labels (usually semiconductor or metal NPs), which in turn is an indication of the concentration of biomolecule under investigation [74]. The NP labels can be introduced either actively (targeted) or passively (non-targeted) into cells under investigation [76]. The incorporation of nanoparticles in cells as labels allows enhanced visualisation of these cells in vivo using a variety of molecular imaging modalities, such as optical imaging, radionuclide imaging (positron emission tomography or PET), magnetic resonance imaging (MRI) and single-photon emission computed tomography (SPECT)) [77, 78].

vi. **Acting as reactant:** In some cases, significant difference is observed in chemical properties of bulk material and nanoparticles made from it because of high surface-to-volume ratio in case of NPs. This enhanced surface energy of NPs provides them amplified chemical activity. As a result, the nanoparticles have found significant applications in novel electrochemical analysis systems as special reactants [74]. For example, the high reactivity of manganese oxide

NPs (react directly with H_2O_2) in comparison with bulk MnO_2 (catalyses decomposition of H_2O_2) has made these NPs an important component of some electrochemical systems [79].

6. Recent advances in nanotechnology-based biosensors

Recently, techniques that are sensitive, selective, and cost-effective are being developed for detecting diseases and underlying medical issues. In this context, biosensors as nano-electroanalytical tools have taken the centre stage. Progress in the health sector enabled by nanotechnology has facilitated in the management of a number of diseases at an early stage [80]. When studying the electrosensing applications of NPs, transfer of electrons between substrate and active site of enzyme (biocatalyst) forms the basis for the functioning of enzymatic biosensors, and this transfer in turn is transduced to produce an electroanalytical signal [81]. Carbon nanostructures, nanotubes, nanorods, ceramic or polymeric matrices, derivatives of graphene, and other functionalised nanoparticles are some of the materials that have been explored and investigated widely by researchers for their electrosensing/ biosensing properties and applications [82–84]. The following recent researches further ascertain the efficiency and remarkable role of NPs in biosensors:

a. Shen et al. [85] showed that the DNA nanostructures can be used in a variety of biomedical applications such as biosensors and cancer therapy as these can interact with small biomolecules and cancer cells due to their specific engineering, unparalleled programmability, and innate biocompatibility and can also serve as nanocarriers for various therapeutic agents.

b. Outstanding physicochemical properties of nanoscaled polyaniline have captured the interest of scientists in the field of medical, as reported by Kazemi et al. [86]. These nanocomposites have shown the ability to immobilise enzymes, nucleic acids and antigen-antibodies on their surface and thus act as bioreceptors or biocatalysts.

c. A polydimethylsiloxane-based nanostructured immunosensor for the investigation of cortisol in human sweat was fabricated by Liu et al. [87].

d. The electrocatalytic and electrochemical features of nickel-cobalt nanoparticles (chemically reduced on functionalised multi-walled carbon nanotubes) were investigated voltametrically against glucose oxidation by Arikan et al. [88].

e. Polyaniline was sequentially electropolymerised on the surface of the functionalised gold electrode followed by electrodeposition of gold nanoparticles to design a nano-biosensor for rapid and ultrasensitive detection of insulin antibodies against diabetes antigens by Farrokhnia at al [89].

f. Electrochemical immunosensing properties of zirconium oxide nanoflowers integrated with quantum dots were investigated by Gupta et al. [90].

g. Recently Farzin et al. [91] have reviewed the biosensing applications of nanoparticles for early and accurate diagnosis of lentivirus HIV that leads to AIDS.

7. Conclusion

An extra advantage of the extensive use of nanoparticles in electrosensing techniques can be owed to their exceptional attributes, like remarkable selectivity and sensitivity, high surface energy, capability to show morphological as well as functional diversity, portability, cost-effectiveness, ease-of-construction and easy-to-operate. Thus, the merger of electrochemistry and nanotechnology has attributed to a wide range of electro (bio) sensing applications for the detection and quantification of chemical as well as biological target molecules. The role of electrode materials in high-performance electrosensing platforms is crucial and evident. Furthermore, when the simple electrode materials are functionalised with nanostructured and/or nanoengineered materials, significant improvement and enhancement in their conductivity, catalytic activity, biocompatibility, amplification of biorecognition events, acceleration of transduction signal, selectivity, specificity and sensitivity has been observed. Almost every passing day, some new innovation is reported in the field of nano-electrosensing, viz. incorporation of CNTs with metallic NPs aids in the development of highly improved nanocomposites for biosensing applications [86], a dramatic enhancement is observed in the electrical conductivity of rGO (reduced graphene oxide) by incorporation of AgNPs [92], nanowires have an exceptional potential as electrosensing probes because of their high surface-to-volume ratios, reproducibility and extraordinary optical, electrical, and magnetic properties [93], and successful designing of dye-sensitised solar cells based on green zinc oxide nanoparticles [94]. The field of nano-electrochemistry seems to expand its horizon continuously, and despite a significant amount of research done in this field, it just seems to be a new start.

Author details

Rajni Bais
Department of Chemistry, J.N.V. University, Jodhpur, India

*Address all correspondence to: rajni.c31@gmail.com

References

[1] Shetti NP, Nayak DS, Reddy KR, Aminabhvi TM. Graphene–clay-based hybrid nanostructures for electrochemical sensors and biosensors. In: Alagarsamy P, Perumal R, editors. Micro and Nano Technologies; Graphene-Based Electrochemical Sensors for Biomolecules. Amsterdam, Netherlands: Elsevier; 2019. pp. 235-274

[2] Pandey S. Advance nanomaterials for biosensors. Biosensors. 2022;**12**:219

[3] Zhu R, Avsievich T, Popov A, Bykov A, Meglinski I. In vivo nano-biosensing element of red blood cell-mediated delivery. Biosensors & Bioelectronics. 2021;**175**:112845

[4] Hanrahan G, Patil DG, Wang J. Electrochemical sensors for environmental monitoring: Design, development and applications. Journal of Environmental Monitoring. 2004;**6**:657-664

[5] Mustafa G, Khan MS, Asif MI, Ullah A, Khan I, Ullah I. Water pollutants and nanosensors. In: Abd-Elsalam KA, Zahid M, editors. Micro and Nano Technologies. Amsterdam, Netherlands: Elsevier; 2021. pp. 105-133

[6] Sengupta P, Khanra K, Chowdhury AR, Datta P. Lab-on-a-chip sensing devices for biomedical applications. In: Pal K, Kraatz HB, Khasnobish A, Bag S, Banerjee I, Kuruganti U, editors. Woodhead Publishing Series in Electronic and Optical Materials, Bioelectronics and Medical Devices. Sawston, Cambridge: Woodhead Publishing; 2019. pp. 47-95. DOI: 10.1016/B978-0-08-102420-1.00004-2

[7] Somashekharappa KK, Rajendrachari S. Sustainable development information management of carbon nanomaterial-based sensors. In: Manjunatha JG, Hussain CM, editors. Carbon Nanomaterials-based Sensors. Amsterdam, Netherlands: Elsevier; 2022. pp. 3-12

[8] Heyrovsky J. The development of polarographic analysis. The Analyst. 1956;**81**:189-192. DOI: 10.1039/AN9568100189

[9] Eggins BR. Chemical Sensors and Biosensors. Vol. 2. Hoboken, New Jersey, US: John Wiley & Sons; 2002

[10] Venton BJ, Cao Q. Fundamentals of fast-scan cyclic voltammetry for dopamine detection. The Analyst. 2020;**145**(4):1158-1168. DOI: 10.1039/C9AN01586H

[11] Dincer C, Bruch R, Rama EC, Abedul MTF, Merkoçi A, Manz A, et al. Disposable sensors in diagnostics, food, and environmental monitoring. Advanced Materials. 2019;**31**:1806739. DOI: 10.1002/adma.201806739

[12] Isildak Ö, Özbek O. Application of potentiometric sensors in real samples. Critical Reviews in Analytical Chemistry. 2021;**51**(3):218-231

[13] Karimi-Maleh H, Orooji Y, Karimi F, Alizadeh M, Baghayeri M, Rouhi J, et al. A critical review on the use of potentiometric based biosensors for biomarkers detection. Biosensors & Bioelectronics. 2021;**184**:113252. DOI: 10.1016/j.bios.2021.113252

[14] Luppa PB, Sokoll LJ, Chan DW. Immunosensors—Principles and applications to clinical chemistry. Clinica Chimica Acta. 2001;**314**:1-26. DOI: 10.1016/s0009-8981(01)00629-5

[15] Thévenot DR, Toth K, Durst RA, Wilson GS. Electrochemical biosensors: Recommended definitions and classification. Biosensors & Bioelectronics. 2001;**16**:121-131. DOI: 10.1016/s0956-5663(01)00115-4

[16] Ryu H, Thompson D, Huang Y, Li B, Lei Y. Electrochemical sensors for nitrogen species: A review. Sensors and Actuators Reports. 2020;**2**:100022. DOI: 10.1016/j.snr.2020.100022

[17] Guo Y, Compton RG. A bespoke reagent free amperometric chloride sensor for drinking water. The Analyst. 2021;**146**:4700-4707. DOI: 10.1039/D1AN00995H

[18] Moldoveanu SC, David V. Short overviews of analytical techniques not containing an independent separation step. In: Moldoveanu SC, David V, editors. Selection of the HPLC Method in Chemical Analysis. Amsterdam, Netherlands: Elsevier; 2017. pp. 31-53

[19] Shah N, Arain MB, Soylak M. Historical background: Milestones in the field of development of analytical instrumentation. In: Soylak M, Yilmaz E, editors. New Generation Green Solvents for Separation and Preconcentration of Organic and Inorganic Species. Amsterdam, Netherlands: Elsevier; 2020. pp. 45-73

[20] Rajendrachari S, Kumaraswamy BE. Biosynthesis of silver nanoparticles using leaves of acacia Melanoxylon and their application as dopamine and hydrogen peroxide sensors. Physical Chemistry Research. 2020;**8**:1-18

[21] Pooja RM, Chowdhury P. Modern applications of quantum dots: Environmentally hazardous metal ion sensing and medical imaging. In: Hussain CM, Kailasa SK, editors. Micro and Nano Technologies, Handbook of Nanomaterials for Sensing Applications. Amsterdam, Netherlands: Elsevier; 2021. pp. 465-503

[22] Lindino CA, Bulhoes LOS. The potentiometric response of chemically modified electrodes. Analytica Chimica Acta. 1996;**334**:317-322

[23] Erduran V, Bekmezci M, Bayat R, Sen F. Functionalized carbon material-based electrochemical sensors for day-to-day applications. In: Hussain CM, Manjunatha JG, editors. Woodhead Publishing Series in Electronic and Optical Materials, Functionalized Nanomaterial-Based Electrochemical Sensors. Sawston, Cambridge: Woodhead Publishing; 2022. pp. 97-111

[24] Celaya-Padilla JM, Romero-González JS, Galvan-Tejada CE, Galvan-Tejada JI, Luna-García H, Arceo-Olague JG, et al. In-vehicle alcohol detection using low-cost sensors and genetic algorithms to aid in the drinking and driving detection. Sensors. 2021;**21**:7752. DOI: 10.3390/s21227752

[25] Westbroek P, Kiekens P. Probes for pH measurement and simultaneous cellulose removal and bleaching of textiles with enzymes. In: Westbroek P, Priniotakis G, Kiekens P, editors. Woodhead Publishing Series in Textiles, Analytical Electrochemistry in Textiles. Sawston, Cambridge: Woodhead Publishing; 2005. pp. 73-91. DOI: 10.1533/9781845690878.2.71

[26] Shinde PV, Saxena M, Singh MK. Recent developments in graphene-based two-dimensional heterostructures for sensing applications. In: Hywel M, Rout CS, Late DJ, editors. Woodhead Publishing Series in Electronic and Optical Materials, Fundamentals and Sensing Applications of 2D Materials. Sawston, Cambridge: Woodhead

Publishing; 2019. pp. 407-436. DOI: 10.1016/B978-0-08-102577-2.00011-7

[27] Christopher FC, Kumar PS, Christopher FJ, Joshiba GJ, Madhesh P. Recent advancements in rapid analysis of pesticides using nano biosensors: A present and future perspective. Journal of Cleaner Production. 2020;**269**:122356. DOI: 10.1016/j.jclepro.2020.122356

[28] Kumar V, Arora K. Trends in nano-inspired biosensors for plants. Materials Science for Energy Technologies. 2020;**3**:255-273. DOI: 10.1016/j.mset.2019.10.004

[29] Shawon ZBZ, Hoque ME, Chowdhury SR. Nanosensors and nanobiosensors: Agricultural and food technology aspects. In: Pal K, Gomes F, editors. Micro and Nano Technologies, Nanofabrication for Smart Nanosensor Applications. Amsterdam, Netherlands: Elsevier; 2020. pp. 135-161

[30] Belluzo MS, Ribone ME, Lagier CM. Assembling Amperometric biosensors for clinical diagnostics. Sensors. 2008;**8**:1366-1399. DOI: 10.3390/s8031366

[31] Wu X, Hou L, Lin X, Xie Z. Application of novel nanomaterials for chemo- and biosensing of algal toxins in shellfish and water. In: Wang X, Chen X, editors. Micro and Nano Technologies, Novel Nanomaterials for Biomedical, Environmental and Energy Applications. Amsterdam, Netherlands: Elsevier; 2019. pp. 353-414

[32] Özbek O, Berkel C, Isildak Ö. Applications of potentiometric sensors for the determination of drug molecules in biological samples. Critical Reviews in Analytical Chemistry. 2022;**52**(4):768-779

[33] Shahdeo D, Gandhi S. Next generation biosensors as a cancer diagnostic tool. In: Khan R, Parihar A, Sanghi SK, editors. Biosensor Based Advanced Cancer Diagnostics. Cambridge, Massachusetts: Academic Press; 2022. pp. 179-196

[34] Regtien P, Dertien E. Optical sensors. In: Regtien P, Dertien E, editors. Sensors for Mechatronics. Amsterdam, Netherlands: Elsevier; 2018. pp. 183-243

[35] Wei H et al. An overview of lead-free piezoelectric materials and devices. Journal of Materials Chemistry C. 2018;**6**:12446-12467

[36] de Rivera PJ, de Rivera MR, Socorro F, de Rivera MR. Calibration and operation improvements of a calorimetric sensor for medical applications. Measurement. 2021;**186**:110134

[37] Holzinger M, Le Goff A, Cosnier S. Nanomaterials for biosensing applications: A review. Frontiers in Chemistry. 2014;**2**:63-81. DOI: 10.3389/fchem.2014.00063

[38] Javaid M, Haleem A, Singh RP, Rab S, Suman R. Exploring the potential of nanosensors: A brief overview. Sensors International. 2021;**2**:100130

[39] Li Y, Schluesener H, Xu S. Gold nanoparticle-based biosensors. Gold Bulletin. 2010;**43**:29-41. DOI: 10.1007/BF03214964

[40] Wijaya E, Lenaerts C, Maricot S, et al. Surface plasmon resonance-based biosensors: From the development of different SPR structures to novel surface functionalization strategies. Current Opinion Solid State Material Sciences. 2011;**15**:208-224. DOI: 10.1016/j.cossms.2011.05.001

[41] Stozhko NY et al. Electrochemical properties of Phytosynthesized gold nanoparticles for Electrosensing. Sensors

(Basel). 2021;**22**:311. DOI: 10.3390/s22010311

[42] Liu CS, Li J, Pang H. Metal-organic framework-based materials as an emerging platform for advanced electrochemical sensing. Coordination Chemistry Reviews. 2020;**410**:213222

[43] Kalambate PK, Noiphung J, Rodthongkum N, Larpant N, Thirabowonkitphithan P, Rojanarata T, et al. Nanomaterials-based electrochemical sensors and biosensors for the detection of non-steroidal anti-inflammatory drugs. TrAC, Trends in Analytical Chemistry. 2021;**143**:116403

[44] Huihui L, Bingfang H, Bingbing G. Emerging electrochemical sensors for life healthcare. Engineered Regeneration. 2021;**2**:175-181

[45] Clarance P, Luvankar B, Sales J, Khusro A, Agastian P, Tack JC, et al. Green synthesis and characterization of gold nanoparticles using endophytic fungi Fusarium solani and its in-vitro anticancer and biomedical applications. Saudi Journal of Biological Sciences. 2020;**27**:706-712. DOI: 10.1016/j.sjbs.2019.12.026

[46] Nejati K, Dadashpour M, Gharibi T, Mellatyar H, Akbarzadeh A. Biomedical applications of functionalized gold nanoparticles: A review. Journal of Cluster Science. 2021;**33**:1-16

[47] Tabatabaei MS, Islam R, Ahmed M. Applications of gold nanoparticles in ELISA, PCR, and immuno-PCR assays: A review. Analytica Chimica Acta. 2021;**1143**:250-266

[48] Yang D, Tian Q, Dong Y, Xing L, Gopinath SCB, Mao Y. Gold-silane complexed antibody immobilization on polystyrene ELISA surface for enhanced determination of matrix Metalloproteinase-9. Process Biochemistry. 2021;**100**:231-236

[49] Jameson DM, James NG, Albanesi JP. Fluorescence fluctuation spectroscopy approaches to the study of receptors in live cells. In: Tetin SY, editor. Methods in Enzymology. Cambridge, Massachusetts: Academic Press; 2013. pp. 87-113. DOI: 10.1016/B978-0-12-405539-1.00003-8

[50] Karlsson HL, Toprak MS, Fadeel B. Toxicity of metal and metal oxide nanoparticles. In: Nordberg GF, Fowler BA, Nordberg M, editors. Handbook on the Toxicology of Metals. 4th ed. Cambridge, Massachusetts: Academic Press; 2015. pp. 75-112

[51] Sohal N, Maity B, Basu S. Recent advances in heteroatom-doped graphene quantum dots for sensing applications. RSC Advances. 2021;**11**:25586-25615. DOI: 10.1039/D1RA04248C

[52] Khose RV, Bangde P, Bondarde MP, Dhumal PS, Bhakare MA, Chakraborty G, et al. Waste derived approach towards wealthy fluorescent N-doped graphene quantum dots for cell imaging and H_2O_2 sensing applications. Spectrochimica Acta Part A: Molecular and Biomolecular Spectroscopy. 2022;**266**:120453

[53] Masteri-Farahani M, Ghorbani F, Mosleh N. Boric acid modified S and N co-doped graphene quantum dots as simple and inexpensive turn-on fluorescent nanosensor for quantification of glucose. Spectrochimica Acta, Part A: Molecular and Biomolecular Spectroscopy. 2021;**245**:118892. DOI: 10.1016/j.saa.2020.118892

[54] HeH YY, Li J, Lai X, Chen X, Wang L, Zhang W, et al. Enhanced fluorescence of Zn-doped carbon quantum dots using zinc citrate chelate as precursor for fluorescent sensor applications.

Materials Science and Engineering B. 2021;**264**:114955

[55] Anusha T, Bhavani KS, Kumar JS, Brahman PK. Synthesis and characterization of novel lanthanum nanoparticles-graphene quantum dots coupled with zeolitic imidazolate framework and its electrochemical sensing application towards vitamin D3 deficiency. Colloids and Surfaces A: Physicochemical and Engineering Aspects. 2021;**611**:125854

[56] Wang J, Ning H, Wang J, Kershaw SV, Jing L, Xiao P. Effects of repetitive pressure on the photoluminescence of bare and ZnS-capped CuInS2 quantum dots: Implications for nanoscale stress sensors. ACS Applied Nano Materials. 2022;**5**:5617-5624. DOI: 10.1021/acsanm.2c00573

[57] Rocha-Santos TAP. Sensors and biosensors based on magnetic nanoparticles. Trends in Analytical Chemistry. 2014;**62**:28-36. DOI: 10.1016/j.trac.2014.06.016

[58] Bishop KJM, Wilmer CE, Soh S, Grzybowski BA. Nanoscale forces and their uses in self-assembly. Small. 2009;**5**:1600-1630. DOI: 10.1002/smll.200900358

[59] Afshar EA, Taher MA, Karimi-Maleh H, Karaman C, Joo SW, Vasseghian Y. Magnetic nanoparticles based on cerium MOF supported on the MWCNT as a fluorescence quenching sensor for determination of 6-mercaptopurine. Environmental Pollution. 2022;**305**:119230. DOI: 10.1016/j.envpol.2022.119230

[60] Rajendrachari S, Karaoglanli AC, Ceylan Y, Uzun O. A fast and robust approach for the green synthesis of spherical magnetite (Fe_3O_4) nanoparticles by Tilia tomentosa (Ihlamur) leaves and its antibacterial studies. Pharmaceutical Sciences. 2020;**26**:175-183. DOI: 10.34172/PS.2020.5

[61] Rajendrachari S, Ceylan KB. The activation energy and antibacterial investigation of spherical Fe3O4 nanoparticles prepared by Crocus sativus (saffron) flowers. Biointerface Research in Applied Chemistry. 2020;**10**:5951-5959. DOI: 10.33263/BRIAC104.951959

[62] Zhu H, Lu Y, Xia J, Liu Y, Chen J, Lee J, et al. Aptamer-assisted protein orientation on silver magnetic nanoparticles: Application to sensitive leukocyte cell-derived Chemotaxin 2 surface plasmon resonance sensors. Analytical Chemistry. 2022;**94**(4):2109-2118. DOI: 10.1021/acs.analchem.1c04448

[63] Rajendrachari S, Gururaj JK, Prakashaiah BG, Kumar M, Swamy BEK. Electrocatalytic determination of ascorbic acid using a green synthesised magnetite nano-flake modified carbon paste electrode by cyclic voltammetric method. Materials Research Innovations. 2022;**26**:229-239

[64] Tran HV, Ngo NM, Medhi R, Srinoi P, Liu T, Rittikulsittichai S, et al. Multifunctional Iron oxide magnetic nanoparticles for biomedical applications: A review. Materials. 2022;**15**:503. DOI: 10.3390/ma15020503

[65] Saxena S, Srivastava AK. Carbon nanotube-based sensors and their application. In: Thomas S, Grohens Y, Vignaud G, Kalarikkal N, James J, editors. Micro and Nano Technologies, Nano-optics. Amsterdam, Netherlands: Elsevier; 2020. pp. 265-291

[66] Zaporotskova IV, Boroznina NP, Parkhomenko YN, Kozhitov LV. Carbon

nanotubes: Sensor properties. A review. Modern Electronic Materials. 2016;**2**:95-105

[67] Schroeder V, Savagatrup S, He M, Lin S, Swager TM. Carbon nanotube chemical sensors. Chemical Reviews. 2019;**119**:599-663. DOI: 10.1021/acs.chemrev.8b00340

[68] Wang Y, Zhang X, Zhou T, Cui Z, Zhang K. A novel terahertz metasurface based on a single-walled carbon nanotube film for sensing application. Journal of Materials Chemistry A. 2022;**10**:1780-1787. DOI: 10.1039/D1TA09396G

[69] Verma P, Ubaid J, Varadarajan KM, Wardle BL, Kumar S. Synthesis and characterization of carbon nanotube-doped thermoplastic nanocomposites for the additive manufacturing of self-sensing piezoresistive materials. ACS Applied Materials & Interfaces. 2022;**14**:8361-8372. DOI: 10.1021/acsami.1c20491

[70] Ahmadian E, Janas D, Eftekhari A, Zare N. Application of carbon nanotubes in sensing/monitoring of pancreas and liver cancer. Chemosphere. 2022;**302**:134826. DOI: 10.1016/j.chemosphere.2022.134826

[71] Zahirinejad S, Hemmati R, Homaei A, Dinari A, Hosseinkhani S, Mohammadi S, et al. Nano-organic supports for enzyme immobilization: Scopes and perspectives. Colloids and Surfaces. B, Biointerfaces. 2021;**204**:111774. DOI: 10.1016/j.colsurfb.2021.111774

[72] Malhotra BD, Ali MA. Nanomaterials in biosensors: Fundamentals and applications. In: Malhotra BD, Ali MA, editors. Nanomaterials for Biosensors. Norwich, New York: William Andrew Publishing; 2018. pp. 1-74

[73] Lin H, Wei K, Yin Z, Sun S. Nanocatalysts in electrosynthesis. iScience. 2021;**24**:102172. DOI: 10.1016/j.isci.2021.102172

[74] Luo X, Morrin A, Killard AJ, Smyth MR. Nanoparticles in electrochemical biosensors. In: Nalwa HS, editor. Encyclopedia of Nanoscience and Nanotechnology. Valencia, California: American Scientific Publishers; 2011. pp. 473-501

[75] Millet P, Ngameni R, Grigoriev SA, Mbemba N, Brisset F, Ranjbari A, et al. PEM water electrolyzers: From electrocatalysis to stack development. International Journal of Hydrogen Energy. 2010;**35**:5043-5052. DOI: 10.1016/j.ijhydene.2009.09.015

[76] Ruoslahti E, Bhatia SN, Sailor MJ. Targeting of drugs and nanoparticles to tumors. The Journal of Cell Biology. 2010;**188**:759-768. DOI: 10.1083/jcb.200910104

[77] Xu C, Xie J, Ho D, Wang C, Kohler N, Walsh EG, et al. Au–Fe3O4 dumbbell nanoparticles as dual-functional probes. Angewandte Chemie, International Edition. 2008;**4**:173-176. DOI: 10.1002/anie.200704392

[78] Liong M, Lu J, Kovochich M, Xia T, Ruehm SG, Nel AE, et al. Multifunctional inorganic nanoparticles for imaging, targeting, and drug delivery. ACS Nano. 2008;**2**:889-896. DOI: 10.1021/nn800072t

[79] Xu JJ, Zhao W, Luo XL, Chen HY. A sensitive biosensor for lactate based on layer-by-layer assembling MnO_2nanoparticles and lactate oxidase on ion-sensitive field-effect transistors. Chemical Communications. 2005;**6**:792-794

[80] Pradhan A, Lahare P, Sinha P, Singh N, Gupta B, Kuca K, et al.

Biosensors as nano-analytical tools for COVID-19 detection. Sensors. 2021;**21**:7823. DOI: 10.3390/s21237823

[81] Colmati F, Sgobbi LF, Teixeira GF, Vilela RS, Martins TD, Figueiredo GO. Electrochemical biosensors containing pure enzymes or crude extracts as enzyme sources for pesticides and phenolic compounds with pharmacological property detection and quantification. In: Rinken T, Kivirand K, editors. Biosensors for Environmental Monitoring. London: Intech Open; 2019

[82] Xie Y, Liu T, Chu Z, Jin W. Recent advances in electrochemical enzymatic biosensors based on regular nanostructured materials. Journal of Electroanalytical Chemistry. 2021;**893**:115328

[83] Hajian A, Ghodsi J, Afraz A, Yurchenko O, Urban G. Nanomolar detection of methylparaben by a cost-effective hemoglobin-based biosensor. Materials Science and Engineering: C. 2016;**69**:122-127. DOI: 10.1016/j.msec.2016.06.061

[84] Rajendrachari S, Ramakrishna D. Functionalized nanomaterial-based electrochemical sensors: A sensitive sensor platform. In: Hussain CM, Manjunatha JG, editors. Woodhead Publishing Series in Electronic and Optical Materials, Functionalized Nanomaterial-based Electrochemical Sensors. Sawston, Cambridge: Woodhead Publishing; 2022. pp. 3-25

[85] Shen L, Wang P, Ke Y. DNA nanotechnology-based biosensors and therapeutics. Advanced Healthcare Materials. 2021;**10**:2002205. DOI: 10.1002/adhm.202002205

[86] Kazemi F, Naghib SM, Zare Y, Rhee KY. Biosensing applications of polyaniline (PANI)-based nanocomposites: A review. Polymer Reviews. 2021;**61**:553-597

[87] Liu Q, Shi W, Tian L, Su M, Jiang M, Li J, et al. Preparation of nanostructured PDMS film as flexible immunosensor for cortisol analysis in human sweat. Analytica Chimica Acta. 2021;**1184**:339010. DOI: 10.1016/j.aca.2021.339010

[88] Arikan K, Burhan H, Bayat R, Sen F. Glucose nano biosensor with non-enzymatic excellent sensitivity prepared with nickel–cobalt nanocomposites on f-MWCNT. Chemosphere. 2022;**291**:132720. DOI: 10.1016/j.chemosphere.2021.132720

[89] Farrokhnia M, Amoabediny G, Ebrahimi M, Ganjali M, Arjmand M. Ultrasensitive early detection of insulin antibody employing novel electrochemical nano-biosensor based on controllable electro-fabrication process. Talanta. 2022;**238**:122947. DOI: 10.1016/j.talanta.2021.122947

[90] Gupta PK, Chauhan D, Khan ZH, Solanki PR. ZrO2 nanoflowers decorated with graphene quantum dots for electrochemical immunosensing. ACS Applied Nano Materials. 2020;**3**:2506-2516. DOI: 10.1021/acsanm.9b02598

[91] Farzin L, Shamsipur M, Samandari L, Sheibani S. HIV biosensors for early diagnosis of infection: The intertwine of nanotechnology with sensing strategies. Talanta. 2020;**206**:120201. DOI: 10.1016/j.talanta.2019.120201

[92] Cho IH et al. Current Technologies of Electrochemical Immunosensors: Perspective on signal amplification. Sensors. 2018;**18**:207

[93] Zeb S, Sun G, Nie Y, Cui Y, Jiang X. Synthesis of highly oriented WO3 nanowire bundles decorated with Au for gas sensing application. Sensors and Actuators B: Chemical.

2020;**321**:128439. DOI: 10.1016/j.snb.2020.128439

[94] Rajendrachari S, Esgin H, Yilmaz VM, Caglar Y. Fabrication and characterization of green synthesized ZnO nanoparticle based dye-sensitized solar cells. Journal of Science: Advanced Materials and Devices. 2020;**5**:185-191. DOI: 10.1016/j.jsamd.2020.04.005

Section 3

Cyclic Voltammetric Applications with Recent Advancements

Chapter 4

A Review on Cyclic Voltammetric Investigation of Toxic Heavy Metals

Shashanka Rajendrachari, Kiran Kenchappa Somashekharappa, Rayappa Shrinivas Mahale, Shamanth Vasanth and Sharath Peramenahalli Chikkegouda

Abstract

Heavy metals are one of the toxic pollutants threatening the human kind by causing various health issues. The detection of such polutants are of important environmental concern and we need a real-time monitoring equipment. Many researchers have established a number of approaches for the detection of these heavy metals so far. But, the development of one time use sensors for the quick, and real time detection of toxic heavy metals is in great demand. The electrochemical methods like cyclic voltammetry, is proved to be one of the best and popular methods, and are preferred over other electrochemical methods because of its high sensitivity, selectivity, anti-fouling, quick and accurate detection. In the present book chapter, we will discuss the various modifiers used to detect the arsenic, cadmium, and lead heavy metals using cyclic voltammetry.

Keywords: heavy metals, electrochemical sensors, cyclic voltammetry, electrodes, toxic pollutants

1. Introduction

Heavy metals occur either naturally by geological activities, or by agricultural and industrial wastes. Any elements exhibiting relatively high molecular weight and more than 5 g/cm^3 density are considered as heavy metals [1]. They are generally non-biodegradable and toxic in nature due to their ability to transform from one oxidation state to another easily. The heavy metals can cause bioaccumulation because of absorption of them by plants and animals living in that contamination areas [2] and causes various health effects like malfunction of gastrointestinal, nervous system, kidney and immune system, followed by birth defects, skin lesions, and cancer [3]. **Figure 1** depicts the different organs of humans affected by the consumption of various heavy metals.

The agricultural and industrial wastes like pesticides, fungicides, refineries, fertilizers, mining, smoking, nuclear fission plants, chemical industry, paint, electroplating, welding, automobiles, batteries, are the major sources of heavy metal ions [5]. Currently, there are more than 50 heavy metals are known and some of them are metalloids, actinides, transition elements and lanthanides [6]. The elements like lead, mercury, arsenic and cadmium are considered as very toxic and their consumption

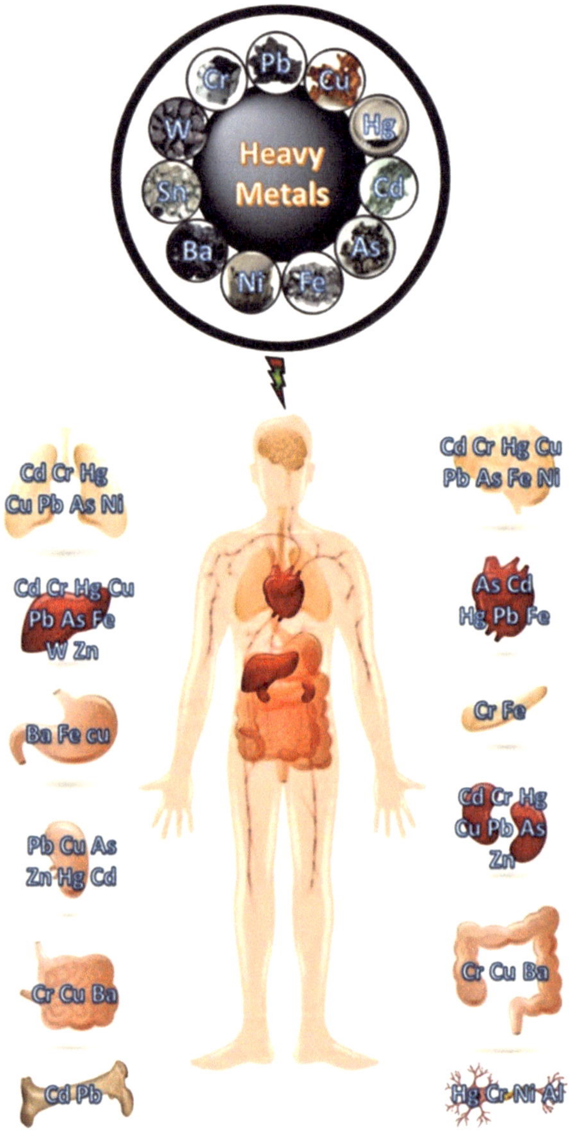

Figure 1.
Various heavy metals affecting the different organs of human [4].

in small amount can cause serious health issues. But, some of the heavy metals are human friendly and their consumption in trace amount can maintain good health. Therefore, regular determination of the heavy metals is advisable in the contamination areas. **Figure 2** depicts the possible ways of heavy metal exposure, their impact on human health and their mechanism.

Many researchers used cold vapor atomic fluorescence, atomic absorption, and emission spectroscopies and inductively coupled plasma techniques [8] to detect heavy metals. But these methods are time consuming, chances of contamination, requires huge manpower and area. Therefore, we need some standard analytical methods to detect the presence of heavy metal ions in foods, plants, animals, water, and soils. Electrochemical methods like cyclic voltammetry [9–16], stripping voltammetry,

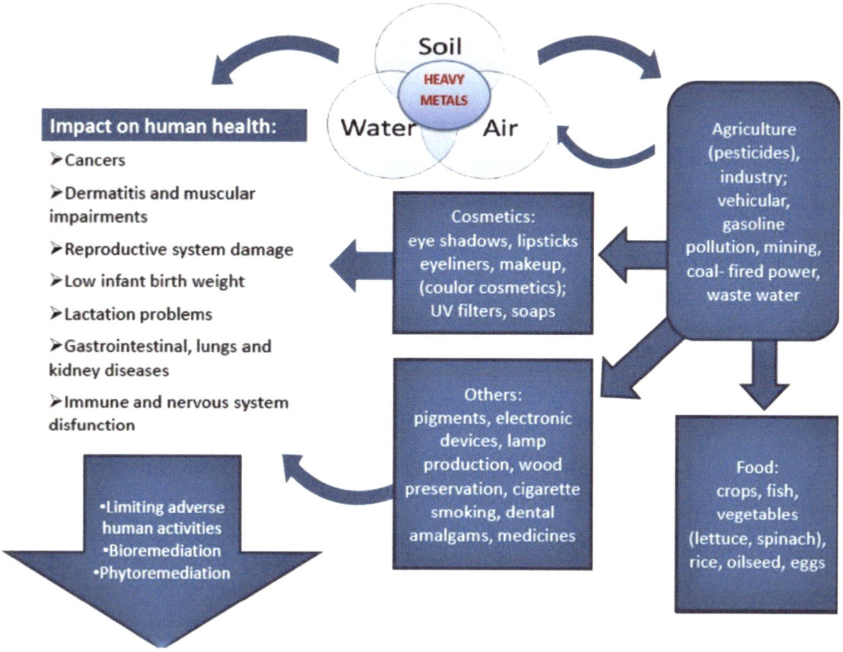

Figure 2.
The possible ways of heavy metal exposure, their impact on human health and their mechanism [7].

differential pulse voltammetry, and polarography are so far best and popular choice to determine the various heavy metals in the environment. These methods are cheap, highly accurate, quick, robust in nature. Among them, cyclic voltammetry is proved to be one of the better and highly advantageous electrochemical methods used to detect organic, inorganic, organometallic, and biological heavy metal ions. High sensitivity, fast response, live monitoring data acquisition, wide detection limit, and possibility of simultaneous detection of multi elements by surface functionalization has made it one of the popular choices among the environmentalists used to detect the heavy metals [6, 17–21]. The present book chapter focus on the recent developments, current challenges, and prospects for future research in cyclic voltammetric determination of lead, arsenic, and cadmium in the environment. We hope that this book chapter will be useful for all the environmental researchers working on heavy metals.

2. Electrochemical determination of Lead (Pb) using cyclic voltammetry

Lead is a bluish-gray metal generally found in the earth crust. Lead is very poisonous metal directly affect the human nervous system and causes severe headaches, and memory loss [22]. Lead is very dangerous woman especially during the pregnancy, when a pregnant woman consumes lead, then it reaches the fetus and causes premature childbirth, abnormal growth, and low weight of the fetus. Lead also affect the brain developments in children and causes the abnormalities [23]. Therefore, people working in lead contamination zone must regularly check the amount of lead in their body.

Mei et al. have reported the use of TC4 arene-modified screen-printed carbon electrode (SPCE) to detect lead ions in river water using cyclic voltammetry [22].

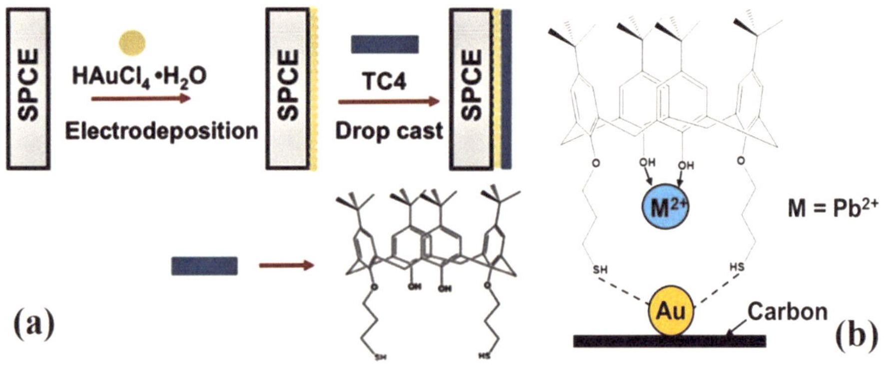

Figure 3.
(a) The schematic representation of electrode preparation to determine Pb2+, and (b) binding of electrode with Pb^{2+} *[22].*

Figure 3 demonstrates the schematic representation of electrode preparation to determine Pb^{2+}. They reported the 0.7982×10^{-2} ppm detection limit for Pb^{2+} followed by the excellent reproducibility and stability. **Figure 4** depicts the current response of the sensor detecting Pb^{2+} ions at different pH levels.

Riyanto has determined lead ions by cyclic voltammetry method using platinum wire as a working electrode in wastewater [24]. Authors reported that, they successfully detected the lead from wastewater with correlation of determination ($R^2 = 0.999$), LOD of 0.9029 mg/L, limit of quantification (LOQ) of 3.0098 mg/L and recovery of 100.67% respectively. They found that, the electro-oxidation of lead on Pt wire electrode occurs in a reversible system and reported that the analytical parameter from cathodic peak is better compared to anodic peak for analyzing of Pb using CV method. They claimed that, the fabricated electrode is simple, economic and showed an excellent selectivity, accuracy, sensitivity, reproducibility [24].

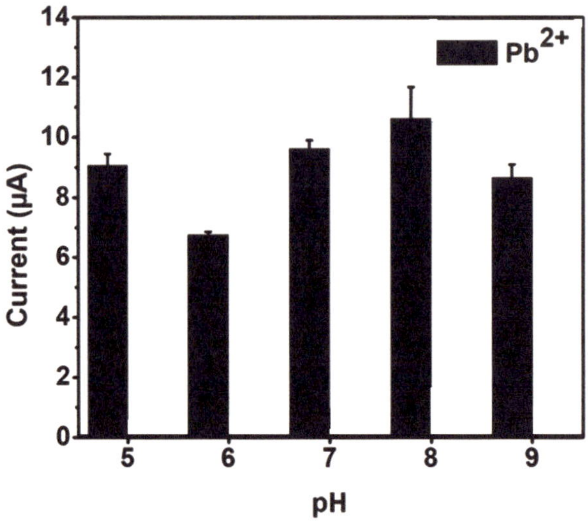

Figure 4.
Current response of electrode during electrochemical determination of lead ions in a KCl supporting electrolyte at different pH [22].

Khodari et al. İnvestigated the cyclic voltammetric behavior of lead ions well water samples using glassy carbon electrode [25]. Cyclic voltammogram showed an anodic peak at −520 mV for lead ions. Authors reported that, the glassy carbon electrode showcased an excellent linear response to a good linear response to Pb^{2+} in the concentration range from 8×10^{-6} M to 1×10^{-4} M with a detection limits of 2×10^{-7} M. They also studied the effect of scan rate, deposition time, deposition potential, and the pH of the supporting electrolyte on lead ions. They claimed that, the fabricated electrode is accurate, precise, highly selective in determining the lead ions in well water.

Honeychurch used carbon rod electrode extracted from zinc-carbon batteries to determine the traces of lead ions present in tap water sample using cyclic voltammetry [26]. Author studied the electrochemical behavior of Pb at different supporting electrolytes like ortho-phosphoric acid, HNO_3, HCl, CH_3COOH, KCl and malonic acid. They found the optimum electrochemical condition of a supporting electrolyte

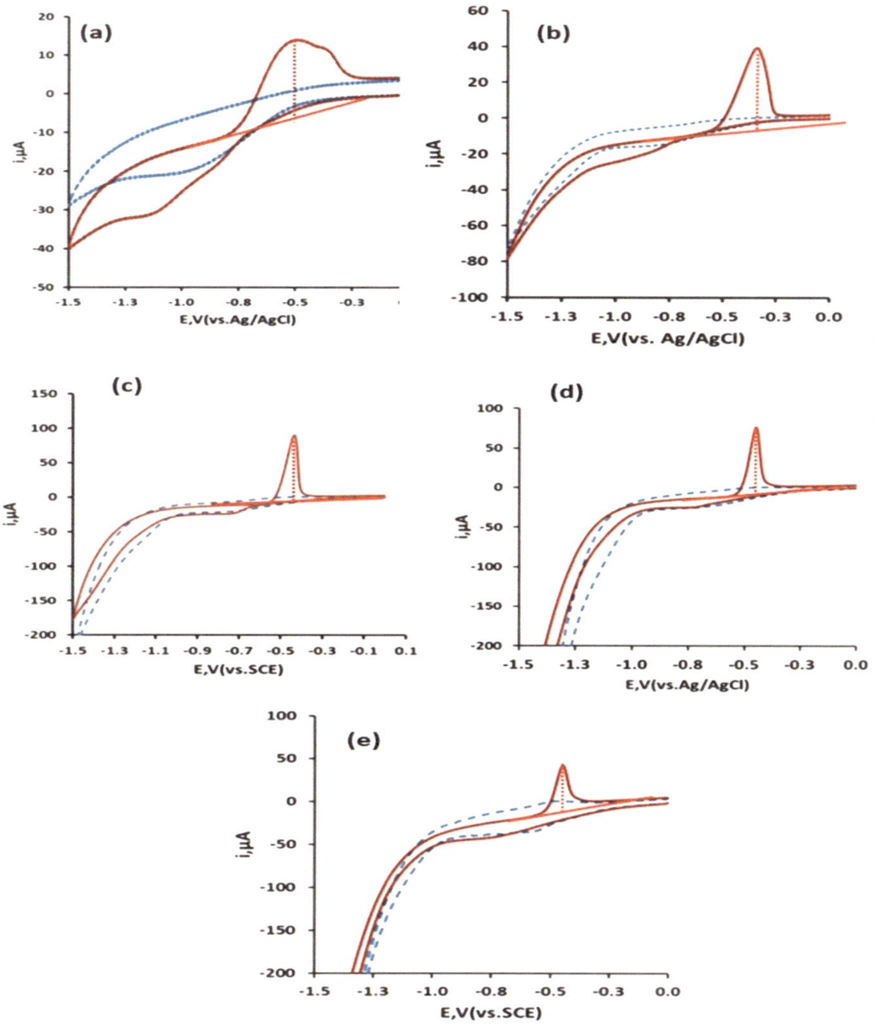

Figure 5.
Effect of acetic acid supporting electrolyte concentration on the cyclic voltammetric behavior of 116 μM Pb: (a) 0.0 M, (b) 0.03 M, (c) 0.1 M, (d): 0.66 M, and (e) 3.0 M [26].

of 4% v/v acetic acid, with a deposition potential and time of −1.5 V (vs. SCE) and 1100 seconds as per the **Figure 5**. They reported the linear range of 2.8 μg/L to 110 μg/L and a detection limit of 2.8 μg/L with 95.6% mean recovery.

3. Electrochemical determination of arsenic (As) using cyclic voltammetry

Arsenic is the 20th most abundant and high mobile element found in earth crust and exists as dust in air and dissolves in rain water and other water sources easily. The World health organization standard for As concentration must be 10 μg/L, but some parts of the World it exceeds the standard concentration. The excessive intake

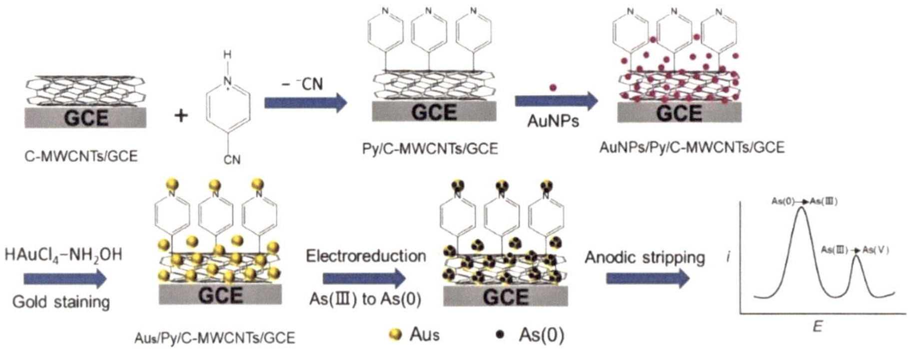

Figure 6.
Preparation of the electrode [28].

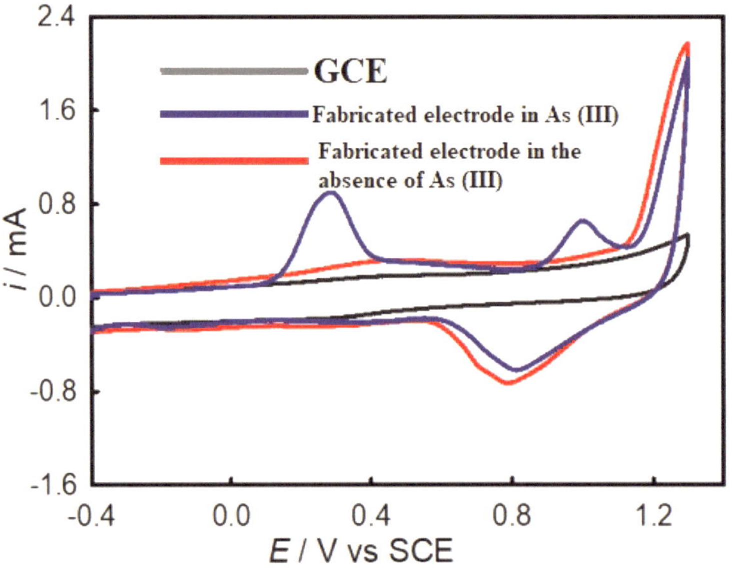

Figure 7.
Cyclic voltammogramms GCE and the fabricated Aus/Py/C-MWCNTs/GCE electrode in 0.1 M aqueous H_2SO_4 with 1 μM As(III) and without As (III) [28].

of As causes lung and skin cancer along with negative impacts on cognitive development and increased deaths [27]. Therefore, detection of arsenic in the environment is important for maintaining public health.

Du et al. fabricated a novel Au-stained Au nanoparticles/pyridine/carboxylated multiwalled carbon nanotubes/glassy carbon electrode for the detection of Arsenic (III) traces in real water samples using cyclic voltammetry [28]. The fabrication of the electrode is schematically represnted in **Figure 6**.

Authors observed the peak currents for oxidation of As (0) to As (III) are linear with a concentration of As (III) from 0.01 to 8 μM with a sensitivity of 0.741 mA μM^{-1} and an LOD of 3.3 nM respectively. Similarly, for the peak currents fort he oxidation of As (III) to As (V) are found to be linear from 0.01 to 8.0 μM with a sensitivity of 0.175 mA μM^{-1} and an LOD of 16.7 nM respectively. **Figure 7** depicts the cyclic voltammogramms of prepared and gassy carbon electrode (GCE) during the electrochemical behavior of As (III).

Ismail et al. fabricated low cost silica nanoparticles modified SPCE for the detection of As (III) using cyclic voltammetry as shown in the **Figure 8** [29]. They performed optimzation of the experimental conditions and reported that, the anodic peak current exhibited a linear range of 5 to 30 μg/L to the As(III) concentration, with a LOD of 6.2 μg/L. Authors claimed that, the fabricated electrode is very economic with high accuracy, high selectivity, sensitivity, stability with good reproducebility.

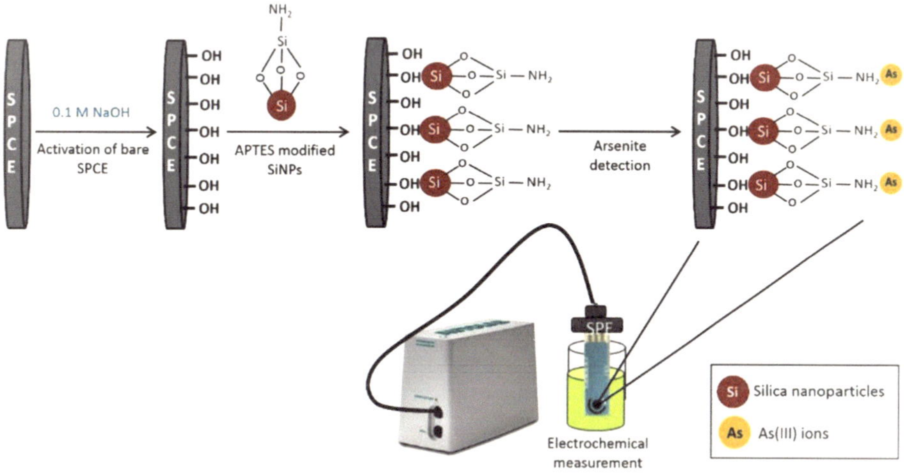

Figure 8.
Fabrication of silica nanoparticles modified SPCE [29].

Modified electrode	Linear range (μg/L)/ppb	LOD (μg/L)/ppb	Reference
Au/SPE	0.5–12	0.22	[30]
Au-Cu/GCE	30–130	5.64	[31]
Si/SPCE	5–30	6.2	[29]
Au/GCE	0.01–8	0.0033	[28]
Au/GCE	0.01–8	0.0167	[28]

Table 1.
Different electrodes used to detect arsenic by cyclic voltammetric method.

Trachioti et al. fabricated a sparked gold nanoparticles from eutectic Au/Si alloy for the determination of arsenic ın drinking water using cyclic voltammetry [30]. They reported that the anodic peak current was proportional to the arsenic concentration over a linear range of 0.5 to 12 5 ppb, with a LOD of 6.2 ppb. The fabricated electrode exhibited excellent detection capability, high selectivity and reproducibility. Authors claimed that, the fabrication method of electrode is extremely simple and economic with wide scope of applicability. **Table 1** depicts the various electrodes used to detect arsenic by cyclic voltammetric method.

4. Electrochemical determination of cadmium (Cd2+) using cyclic voltammetry

Cadmium ion (Cd^{2+}) are one of most toxic heavy metals that contaminates water and causes deadliest diseases like cancer, kidney dysfunction, cardiovascular disease, bone degeneration, lung, and liver damage to humans [32]. Cadmium ions are considered to have a greater solubility in water than any other heavy metals and offering a great threat to the biosystem. The standard level of Cd^{2+} in drinking water is 0.003 mg/L as per the World Health Organization (WHO). Therefore, the detection of Cd^{2+} in drinking water samples is of utmost important.

Attaallah and Amine reported the use of enzymic membrane as a electrode without any pre-treatment to detect cadmium ions in drinking water using cyclic voltammetric method [32]. **Figure 9** depicts the fabrication of enzymic membrane electrode to detect cadmium ions. They also modified the fabricated electrode with screen printed electrodes to further increase the selectivity, sensitivity, reproducibility. They modified electrode has showed the linear calibration range between 0.02–100 ppb (R^2 = 0.990) and a detection limit of 50 ppt respectively. **Figure 10a** depicts the cyclic voltammogram of TMB/H_2O_2, HRP/TMB/H_2O_2, and HRP/Cd^{2+}/TMB/ H_2O_2. [TMB = 3,3′,5,5′-Tetram ethylbenzidine, HRP = horseradish peroxidase] and **Figure 10b** represents the cyclic voltammograms of variation in the cadmium ions from 0.02 to 100 ppb, in 0.1 M acetate buffer.

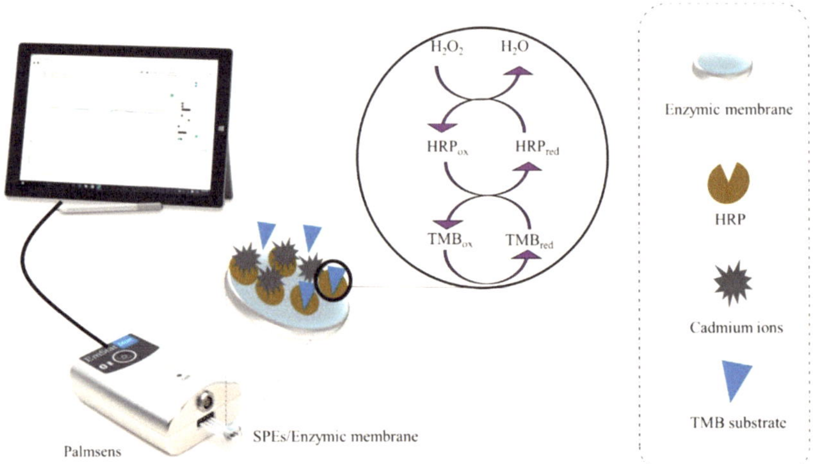

Figure 9.
Schematic representation of preparing enzymic membrane electrode to detect cadmium ions [32].

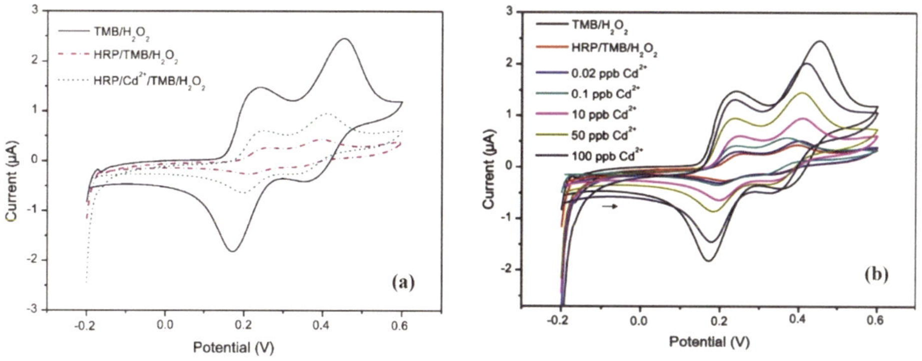

Figure 10.
(a) Cyclic voltammogram of TMB/H_2O_2, HRP/TMB/H_2O_2, and HRP/Cd^{2+}/TMB/ H_2O_2, and (b) cyclic voltammograms of variation in the cadmium ions from 0.02 to 100 ppb, in 0.1 M acetate buffer [32].

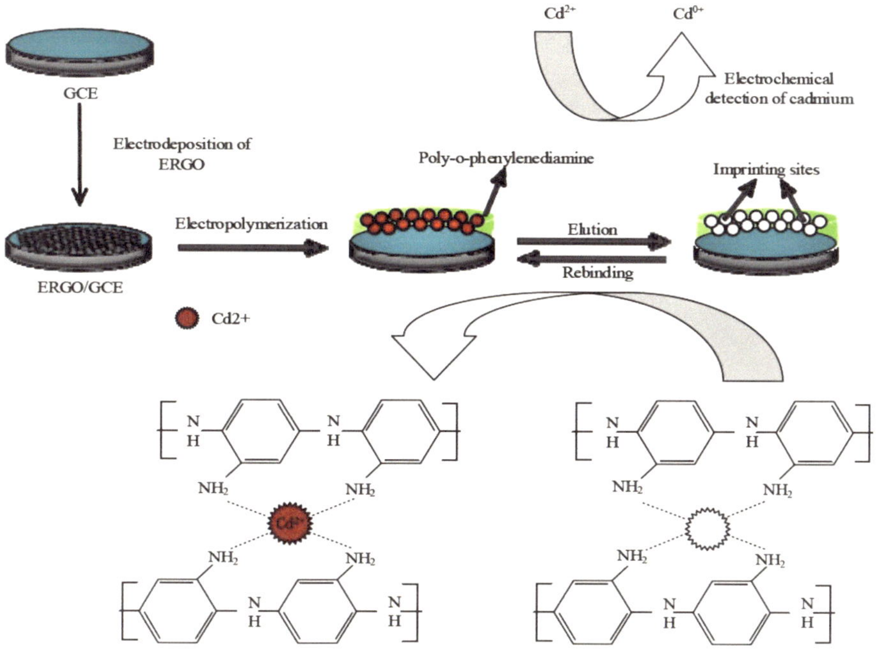

Figure 11.
The procedure diagrams for fabrication of the electrode [33].

Wang et al. electropolymerized ion imprinted poly (o-phenylenediamine) PoPD/ electrochemical reduced graphene (ERGO) composite on glass carbon electrode (GCE) to detect Cd^{2+} in water using cyclic voltammetry [33]. **Figure 11** depicts the procedure diagrams for fabrication of the electrode.

The prepared electrode exhibited a excellent selectivity toward the target Cd(II) ions in the presence multi heavy metal ions. Under optimized conditions, the electrochemical sensor showed a good linear relationship between Cd (II) concentration in the range of 1 to 50 ng/mL, with the limit of detection of 0.13 ng/mL respectively. **Figure 12** depicts the cyclic voltammetric curves of bare GCE, ERGO/GCE, Cd(II)-IIP/ERGO/GCE and IIP/ERGO/GCE.

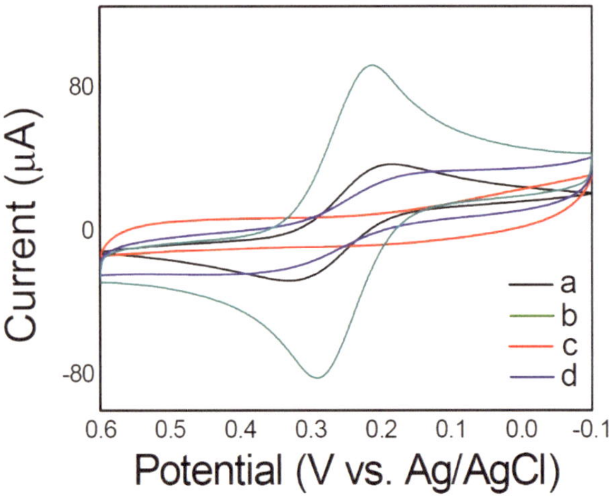

Figure 12.
CV curves on bare GCE (curve a), ERGO/GCE (curve b), Cd(II)-IIP/ERGO/GCE (curve c), and IIP/ERGO/GCE (curve d) between −0.2 V and 0.6 V at a scan rate of 50 mV/s [33].

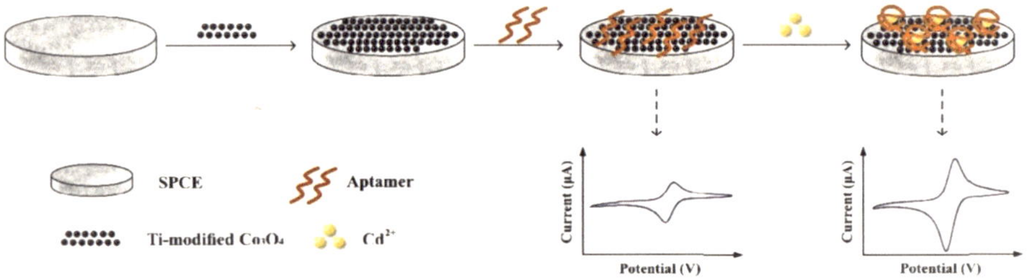

Figure 13.
Schematic representation of fabricting the electrochemical aptasensor used to determine Cd^{2+} [34].

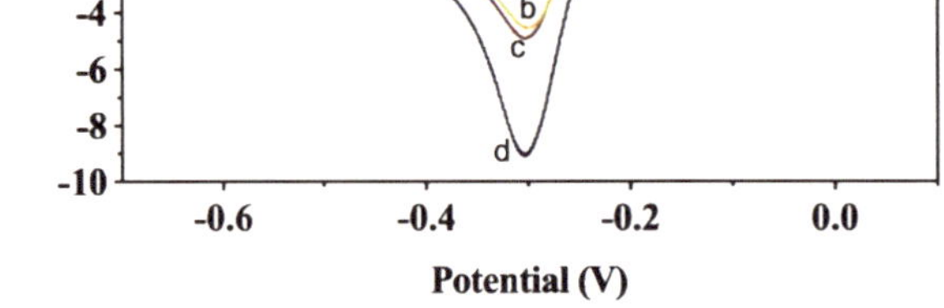

Figure 14.
The CV of (a) bare electrode, (b) Co_2Ti_1 modified electrode, (c) Co_2Ti_1/aptamer modified electrode, and (d) Co_2Ti_1/aptamer/Cd^{2+} modified electrode respectively at 100 mV/s scan rate [34].

Liu et al. prepared Ti-modified Co_3O_4-based electrochemical aptasensor to detect Cd (II) as shown in **Figure 13** [34]. Authors studied the effect of aptamer concentration, incubation time and pH of the solution to optimize the experimental condition to get good results. Under these conditions, peak current was proportional to the Cd (II) concentration over a wide linear range of 0.20 to 15 ng/mL, with a detection limit of 0.49 ng/mL respectively. Authors reported that, they have used cyclic voltammetry method not only to characterize each preparation and optimization step, but also to profile the bindings of aptamer to Cd^{2+}. **Figure 14** depicts the CV of bare electrode, Co_2Ti_1 modified electrode, Co_2Ti_1/aptamer modified electrode, and Co_2Ti_1/aptamer/Cd^{2+} modified electrode respectively.

5. Conclusion

The heavy metals like cadmium, arsenic and lead results in the bioaccumulation and causes adverse effect on gastrointestinal, nervous system, kidney and immune system, and also causes cancer. The detection of heavy metals and their ions is of utmost important in this world. As we discussed, cyclic voltammetry is one of the potential electrochemical method used to detect the heavy metals and their ions easily and quickly. The discussed method is one of the better method and exhibit excellent sensitivity, selectivity, good current response, and possible detection of multi heavy metals simultaneously compared to other electrochemical methods. The recent developments, challenges and future prospectus of the electrochemical detection of heavy metals by cyclic voltammetry is successfully discussed in this chapter. The electrochemical methods are more accurate, simple, robust methods compared to other traditional methods.

Author details

Shashanka Rajendrachari[1*], Kiran Kenchappa Somashekharappa[2], Rayappa Shrinivas Mahale[3,4], Shamanth Vasanth[3] and Sharath Peramenahalli Chikkegouda[5]

1 Department of Metallurgical and Materials Engineering, Bartin University, Bartin, Turkey

2 Department of Chemistry, Govt. First Grade College, Shivamogga, Karnataka, India

3 School of Mechanical Engineering, REVA University, Bengaluru, Karnataka, India

4 Department of Mechanical Engineering, Jain College of Engineering and Research, Belagavi, Karnataka, India

5 Department of Metallurgical and Materials Engineering JAIN Deemed to be University, Bengaluru, Karnataka, India

*Address all correspondence to: shashankaic@gmail.com

References

[1] Katiyar R, Usha Rani KR, Sindhu TS, Sneha Jain HD, Vidhyashree Ashoka VS, Channegowda M. Design and development of electrochemical potentiostat circuit for the sensing of toxic cadmium and lead ions in soil. Engineering Research Express. 2021;**3**:045026. DOI: 10.1088/2631-8695/ac3637

[2] Ding R, Cheong YH, Ahamed A, Lisak G. Heavy metals detection with paper-based electrochemical sensors. Analytical Chemistry. 2021;**93**(4):1880-1888. DOI: 10.1021/acs.analchem.0c04247

[3] Balali-Mood M, Naseri K, Tahergorabi Z, Khazdair MR, Sadeghi M. Toxic mechanisms of five heavy metals: Mercury, Lead, chromium, cadmium, andArsenic. Frontiers in Pharmacology. 2021;**12**:643972. DOI: 10.3389/fphar.2021.643972

[4] Perumal S, Atchudan R, Edison TNJI, Babu RS, Karpagavinayagam P, Vedhi CA. Short review on recent advances of hydrogel-based adsorbents for heavy metal ions. Metals. 2021;**11**:864. DOI: 10.3390/met11060864

[5] Dvoynenko O, Lo S-L, Chen Y-J, Chen GW, Tsai H-M, Wang Y-L, et al. Speciation analysis of Cr(VI) and Cr(III) in water with surface-enhanced Raman spectroscopy. ACS. Omega. 2021;**6**(3):2052-2059. DOI: 10.1021/acsomega.0c05020

[6] Pujol L, Evrard D, Groenen-Serrano K, Freyssinier M, Ruffien-Cizsak A, Gros P. Electrochemical sensors and devices for heavy metals assay in water: The French groups' contribution. Frontiers in Chemistry. 2014;**2**:19. DOI: 10.3389/fchem.2014.00019

[7] Witkowska D, Słowik J, Chilicka K. Heavy metals and human health: Possible exposure pathways and the competition for protein binding sites. Molecules. 2021;**26**:6060. DOI: 10.3390/molecules26196060

[8] Furness RW, Rainbow PS, editors. Heavy Metals in the Marine Environment. Boca Raton, USA: CRC Press; 2018. DOI: 10.1201/9781351073158

[9] Shashanka R, Jayaprakash GK, Prakashaiah BG, Kumar M, Kumara Swamy BE. Electrocatalytic determination of ascorbic acid using a green synthesised magnetite nanoflake modified carbon paste electrode by cyclic voltammetric method. Materials Research Innovations. 2021;**26**:229-239. DOI: 10.1080/14328917.2021.1945795

[10] Mahale RS, Shashanka R, Shamanth V, Vinaykumar R. Voltammetric determination of various food azo dyes using different modified carbon paste electrodes. Biointerface Research in Applied Chemistry. 2022;**12**(4):4557-4566

[11] Mahale RS, Vasanth S, Krishna H, Shashanka R, Sharath PC, Sreekanth NV. Electrochemical sensor applications of nanoparticle modified carbon paste electrodes to detect various neurotransmitters: A review. Applied Mechanics and Materials. 2022;**908**:69-88

[12] Shashanka R, Dileep R. Chapter 1: Functionalized nanomaterial-based electrochemical sensors: A sensitive sensor platform. In: Hussain C, Manjunatha JG, editors. Functionalized Nanomaterial-Based Electrochemical Sensors. Cambridge, England: Elsevier, Woodhead Publishing; 2022. Paperback ISBN: 9780128237885, eBook ISBN: 9780128241851

[13] Kiran KS, Shashanka R. Sustainable development information management

of carbon nanomaterials-based sensors. In: Hussain C, Manjunatha JG, editors. Carbon Nanomaterials-Based Sensors. Amsterdam, Netherlands: Elsevier; 2022. DOI: 10.1016/B978-0-323-91174-0.00001-9. ISBN: 978-0-323-91174-0

[14] Shashanka R, Kumara Swamy BE. Simultaneous electro-generation and electro-deposition of copper oxide nanoparticles on glassy carbon electrode and its sensor application. SN Applied Sciences. 2020;**2**(5):956

[15] Shashanka R, Kumara Swamy BE. Biosynthesis of silver nanoparticles using leaves of Acacia melanoxylon and its application as dopamine and hydrogen peroxide sensors. Physical Chemistry Research. 2020;**8**(1):1-18

[16] Shashanka R, Chaira D, Kumara Swamy BE. Fabrication of yttria dispersed duplex stainless steel electrode to determine dopamine, ascorbic and uric acid electrochemically by using cyclic voltammetry. International Journal of Scientific & Engineering Research. 2016;**7**:1275-1285

[17] Manjunatha JG, Kumara Swamy BE, Mamatha GP, Shankar SS, Gilbert O, Chandrashekar BN, et al. Electrochemical response of dopamine at phthalic acid and triton X-100 modified carbon paste electrode: A cyclic voltammetry study. International Journal of Electrochemical Science. 2009;**4**:1469-1478

[18] Hareesha N, Manjunatha JG, Raril C, Tigari G. Sensitive and selective electrochemical resolution of tyrosine with ascorbic acid through the development of electropolymerized alizarin sodium sulfonate modified carbon nanotube paste electrodes. ChemistrySelect. 2019;**4**:4559-4567

[19] Hareesha N, Manjunatha JG. Surfactant and polymer layered carbon composite electrochemical sensor for the analysis of estriol with ciprofloxacin. Materials Research Innovations. 2020;**24**:349-362

[20] Manjunatha JG. A surfactant enhanced graphene paste electrode as an effective electrochemical sensor for the sensitive and simultaneous determination of catechol and resorcinol. Chemical Data Collections. 2020;**25**:100331

[21] Rajendrachari S, Jayaprakash KG, Pandit A, Karaoglanli AC, Uzun O. Electrocatalytic investigation by improving the charge kinetics between carbon electrodes and dopamine using bio-synthesized CuO nanoparticles. Catalysts. 2022;**12**:994

[22] Mei CJ, Yusof NA, Alang Ahmad SA. Electrochemical determination of Lead & copper ions using Thiolated calix[4]arene-modified screen-printed carbon electrode. Chem. 2021;**9**:157. DOI: 10.3390/chemosensors9070157

[23] Rehman K, Fatima F, Waheed I, Akash MSH. Prevalence of exposure of heavy metals and their impact on health consequences. Journal of Cellular Biochemistry. 2018;**119**:157-184

[24] Riyanto. Determination of lead in waste water using cyclic voltammetry by platinum wire electrode. EKSAKTA;**14**(2):22-33

[25] Khodari M, Ekram M, Rabie M, Fathy M, Mohamed E, Hassan A. Voltammetric determination of cadmium, copper and Lead using glassy carbon electrode. IOSR. Journal of Applied Chemistry. 2014;**7**(9):62-67

[26] Honeychurch K. Trace Voltammetric determination of Lead at a recycled battery carbon rod electrode. Sensors. 2019;**19**:770. DOI: 10.3390/s19040770

[27] Takemura K, Iwasaki W, Morita N, Ohmagari S. High-density and monodisperse electrochemical gold nanoparticle synthesis utilizing the properties of boron-doped diamond electrodes. Nanomaterials. 2022;**12**:1741. DOI: 10.3390/nano12101741

[28] Du Y, Sun C, Shen Y, Liu L, Chen M, Xie Q, et al. Anodic stripping voltammetric analysis of trace arsenic(III) on a Au-stained Au nanoparticles/pyridine/carboxylated multiwalled carbon nanotubes/glassy carbon electrode. Nanomaterials. 2022;**12**:1450. DOI: 10.3390/nano12091450

[29] Ismail S, Yusof NA, Abdullah J, Abd Rahman SF. Electrochemical detection of Arsenite using a silica nanoparticles-modified screen-printed carbon electrode. Materials. 2020;**13**:3168. DOI: 10.3390/ma13143168

[30] Trachioti MG, Karantzalis AE, Hrbac J, Prodromidis MI. Low-cost screen-printed sensors on-demand: Instantly prepared sparked gold nanoparticles from eutectic Au/Si alloy for the determination of arsenic at the sub-ppb level. Sensors & Actuators: B. Chemical. 2019;**281**:273-280. DOI: 10.1016/j.snb.2018.10.112

[31] Yang M, Guo Z, Li LN, Huang YY, Liu JH, Zhou Q, et al. Electrochemical determination of arsenic(III) with ultra-high anti-interference performance using Au–Cu bimetallic nanoparticles. Sensors & Actuators: B. Chemical. 2016;**231**:70-78

[32] Attaallah R, Amine A. An ultrasensitive and selective determination of cadmium ions at ppt level using an Enzymic membrane with colorimetric and electrochemical detection. Biosensors. 2022;**12**:310. DOI: 10.3390/bios12050310

[33] Wang J, Hu J, Hu S, Gao G, Song YA. Novel electrochemical sensor based on Electropolymerized ion imprinted PoPD/ERGO composite for trace Cd(II) determination in water. Sensors. 2020;**20**:1004. DOI: 10.3390/s20041004

[34] Liu Y, Zhang D, Ding J, Hayat K, Yang X, Zhan X, et al. Label-free and sensitive determination of cadmium ions using a Ti-modified Co3O4-based electrochemical Aptasensor. Biosensors. 2020;**10**:195. DOI: 10.3390/bios10120195

Chapter 5

Recent Advances in Voltammetric Sensing

Harsha Devnani and Chetna Sharma

Abstract

The practical day to day life is largely affected by the products that we use, the air that we breathe, the soil that is used to grow crops, the water we drink and use for various household chores or industrial purposes. The purity analysis of these products or estimation of useful inorganic and organic analytes is of utmost importance for avoiding health and environment risk. Everyone wants to be aware that what they are eating or applying on their skin is safe for them. A diabetic patient needs to monitor their blood sugar levels constantly. The air, water and soil quality needs constant monitoring to avoid health hazards. Not just this, chemical analysis is crucial as a crime investigation technique to identify suspects. Fuel quality and storage needs to be tested for eliminating unwanted losses. The electrochemical techniques are inherently fast, selective and sensitive and some systems are portable as well which is a boon for on-site monitoring. Voltammetric techniques like cyclic voltammetry, stripping voltammetry, impedance spectroscopy, amperometry and other techniques provide specific information of the analytes to be tested. This chapter will highlight the voltammetric techniques used for different types of analyte sensing and the advances that have taken place recently related to voltammetric sensing.

Keywords: voltammetry, sensing, electrochemistry, cyclic voltammetry, stripping voltammetry

1. Introduction

The electrochemical sensor is a broad integrated area encompassing physical aspects, analytical science, material science, electronic fabrication, biochemistry along with statistical analysis. This chapter restricts to the electroanalytical methods involving analytical science and electrochemistry which lay the foundation for voltammetric sensing. Electroanalytical measurements are based on the measurement of potential (potentiometry), current (voltammetry) and the amount of electricity consumed or the matter transformed during electrolysis (coulometry). These techniques are inherently fast, sensitive, selective and offer low detection and quantification limits [1]. In contrast to other analytical techniques like chromatography, ICP-OES, MS and others, electrochemical analysis does not require heavy instrumentation and tedious sample preparation. In fact, miniaturized sensors are portable and are thus handy for on-site monitoring wherever required.

Each of the above mentioned electroanalytical measurement methods involve a specifically designed electrochemical cell. Although potentiometric sensors are more lucrative for on-site operations, voltammetric sensors are more sensitive and fast. Recently, there has been a surge in researching new underlying principles for electrochemical sensing. As a result, sensors are now being developed taking advantage of changes in ionic conductivity, resistivity and impedance [2]. An altogether new concept based on imaging of local electrochemical current utilizing the optical signal from the electrode surface (surface plasmon resonance) has been reported by Shan et al. in 2012 [3]. This derives foundation from the fact that the current density can be determined from the local surface plasmon resonance signal.

Electroanalytical techniques like linear sweep voltammetry (LSV), cyclic voltammetry (CV), pulse voltammetry (PV), stripping voltammetry (SV), chronoamperometry provide an in-depth qualitative and quantitative information for an electro-active species deeming them to be a potential alternative to more commonly used spectrometric or chromatographic techniques. The understanding of an electrode process helps to explore mechanisms for in vivo studies, which is beneficial in analyzing how a drug works when administered to a human body. It is being utilized to speciate and determine ultra trace analytes in complex clinical and industrial samples. Sensor development is a thrust area in the field of chemistry, biology and environmental sciences [4]. A chemical sensor is defined as, "a small device that as the result of a chemical interaction or process between the analyte and the sensor device, transforms chemical or biochemical information of a quantitative or qualitative type into an analytically useful signal" [5]. Chemical sensors find wide applications in industry, critical care, quality control, process controls, pollutant monitoring, safety alarms, clinical diagnostics, food adulteration testing, in agriculture and forensics.

With the advent of electrochemistry, initially the voltammetric methods posed a lot of difficulties as a result of which they could not be much benefited from at the time. Gradually, with significant advancements over a period of time, voltammetric sensing soon became popular as an attractive analytical method. All the voltammetric techniques fundamentally involve the application of an electrode potential (E) and measuring the corresponding current (i) in the respective electrochemical cell. In most voltammetric techniques, the applied potential (E) is varied and the resulting current (i) is measured over a period of time (t). The applied potential relates to the change in electroactive species which can undergo a redox reaction and thus there is change in its concentration, which can be related to the corresponding current of the cell [6].

The voltammetric sensors have lured the researchers as they offer high sensitivity for inorganic as well as organic species along with a wide linearity range using a variety of electrolytes over a wide range of temperatures. Moreover, it has a fast response time and offers simultaneous determination of analytes at times. Voltammetric studies extend a deep insight into the kinetics and mechanism of the electrochemical process under study, making them attractive for sensing applications. Stripping analysis by far is the most sensitive voltammetric technique which boasts the advantages that it does not require derivatization and is also less sensitive to matrix changes in comparison to other analytical techniques [7, 8]. It is being widely used for analysis of heavy metals present in trace amounts in environmental samples. Cathodic stripping voltammetry (CSV) and anodic stripping voltammetry (ASV) have also been used for achieving sensitive determination of various analytes like nucleotides, nucleosides and nucleobases [9].

2. Voltammetric sensors

The voltammetric sensors initially were developed from mercury, carbon materials and inert metals as working electrodes. The use of mercury has long been forbidden due to its toxic nature and the noble metals demand high cost. The mercury electrode is almost out of the picture due to its difficult handling and limited anodic potential range. The carbon electrodes are still in use in a number of forms viz. glassy carbon, diamond, fullerene, graphite, nanotubes, graphene and graphene oxide [2]. In the past few years, carbon paste electrodes have gained momentum as working electrodes as they can be screen printed for production at mass scale. Another set of electrodes is the screen printed electrodes which are used as a robust miniaturized version and are viable for commercial technology. Most of the electrochemical sensors in the market, like the glucose sensor, employ screen printed electrodes as the working electrode [10, 11]. There is immense potential for electrochemical sensors to be established in markets other than that of glucose. The recent COVID-19 pandemic has demanded an urgent need for fast diagnostic processes which can help prevent spread of infectious diseases and provide timely diagnosis of neurodegenerative disease which is a crucial factor [12–14]. The voltammetric sensors hold promise for such diagnostic methods as they are fast, simple in operation, offer real time analysis and can be mass-produced for portable use as they can be miniaturized compared to other existing diagnostic methods [15–17].

The research is now targeted in the direction of new electrode materials which could offer comparable sensitivity and cost economical. Carbon and gold based electrodes are more popular owing to their biocompatibility, stability and good electron transfer kinetics. The bare electrodes often lack sensitivity and selectivity needed for the analysis due to poor charge transfer at the electrode surface. This requires modification of electrode surfaces to enhance charge transfer kinetics and avoid interferences [18, 19]. In this light, the nanomaterials are paving the way for smart electrochemical sensors as they provide enhanced electrochemical surface area with large surface area to volume ratio allowing improved interfacial kinetics exhibiting electrocatalytic activity for sensitive and selective determination of analytes [18–20]. The synergistic effect of nanomaterials is also being actively pursued by combining two or more nanomaterials to form a composite which would derive advantages from all individual components based on the sensing need [7, 8, 21–24]. Although nanomaterials have proved to be beneficial, there are certain challenges yet to overcome like handling sophisticated instruments and the stability of nanoparticles utilizing stabilizers or capping agents in certain cases to avoid agglomeration which will disrupt electrode-electrolyte interfacial kinetics [25, 26].

There are mainly three major hurdles encountered in the development of electrochemical sensors: the attainment of a low limit of detection (LOD); limiting the interaction of unwanted interfering species; maintaining sensor stability and achieving reproducibility in complex real matrices [27]. The LOD indicates the lowest concentration/quantity that could be detected for an analyte and is a major criterion for a sensor performance as many times we are dealing with analytes that are present at trace levels in real samples. With the advancement of nanomaterial modified surfaces, picomolar level of detection has been achieved [28–30]. These modified surfaces still pose the challenge of stability and reproducibility. Moreover, the sensor needs to be validated for real samples otherwise it does not hold importance in the market. The test of an electrochemical sensor as a diagnostic tool is only validated if it is stable and functional in a real matrix [31–33]. Real matrices involve interferences that hinder electrode performance. In the area of medical applications, this is being dealt with by exploring advanced materials to improve electrode surfaces. While the passive methods involve the use of polymers to create a hydrophilic and non-charged layer to limit protein absorption, active methods aim to develop stronger shear forces than the adhesion forces of the bound interferents on the surface [34, 35]. Recently, the use of sol–gel materials along with ceramics and nanomaterials for electrode surfaces have proven to enhance the stability of the sensor [36, 37]. The research has progressed over the years to overcome these challenges and establish effective diagnostic tools in the market for various applications. This chapter highlights the advancement that has taken place in recent years in the development of voltammetric sensors for a variety of applications.

2.1 Voltammetric food sensors

Food is prone to contamination of heavy metals, pesticides and adulterants. Metals come from a variety of natural and anthropogenic sources. However, man-made activities such as urbanization and agriculture can increase their levels, pollute water and soil, and damage the environment. Moreover, pesticides and food additives when added beyond their permissible limits are making way through the food chain, affecting human health invariably [38]. Ingestion of dangerous metal-rich vegetables, crustaceans and other foods can damage stable cells, alter metabolism, lead to carcinogenic mutations and toxicological effects on human organs [39]. The consequences of these elements on human health have inspired extensive analysis of major contaminants in food samples. The well-established spectrometric techniques viz. atomic emission or absorption spectroscopy [40], mass spectroscopy (MS) and inductively coupled plasma-MS (ICP-MS) have been widely used for this analysis in the past decade. Locatelli and Melucci (2013) detected levels of copper, mercury, lead, zinc and cadmium in several vegetables using a rectangular ASV approach and a standard addition method. The study was performed using two electrodes, gold electrode for detection of mercury and the other made of mercury to determine lead, copper, zinc and cadmium. For these studies, lettuce, spinach, and tomatoes were differentiated as washed and unwashed categories. The two groups were mineralized after treatment with an acid mixture in different amounts [41, 42]. The investigation showed that the values for lead and cadmium exceed the regulatory limits of the European Council. **Table 1** lists the range of voltammetric techniques used recently to sense a variety of heavy metals with nanomolar detection limits being achieved at robust electrode systems.

Zabihpour et al. (2020) have reported a robust vanillin (flavoring agent) electrochemical sensor derived from carbon paste electrode (CPE) substrate enriched with $NiFe_2O_4$ nanoparticles and 1-hexyl-3-methylimidazolium chloride (1H3MCl). $NiFe_2O_4$ nanoparticles were formed using a co-precipitation approach and the characterization findings pointed to a spherical $NiFe_2O_4$ nanoparticle with a diameter of 22 nm [52]. The CV technique was used to measure the vanillin oxidation peaks at potentials of +690 and + 650 mV on the surface of CPE and $NiFe_2O_4$/1H3MCl/CPE. The DPV analysis at $NiFe_2O_4$/1H3MCl/CPE showed a strong electrocatalytic capacity towards the electrooxidation of vanillin and revealed two distinct oxidation signals at potentials of 640 and 1050 mV. $NiFe_2O_4$/1H3MCl/CPE was thus successfully used as an analytical sensor to determine vanillin and tryptophan levels in coffee, milk, chocolate and cookie samples [52, 53].

It is also required to estimate nutraceuticals or phytopotentials of certain plants for therapeutic purposes. Voltammetry provides a gateway to such analysis with high sensitivity and selectivity. Zheng et al. (2022) have reviewed the evaluation of

Sample	Target	Technique	Electrode	LOD	Reference
Tomato, potato, and mango	Cu, Cd, and Pb	CV and DPSV	PGE modified with MWCNTs	1.03 μg l^{-1} for Cd, 2.12 μg l^{-1} for Cu, and 1.62 μg l^{-1} for Pb	[43]
Fish liver	Cu	SWV	Hanging mercury drop electrode	0.6 nmol l^{-1} and 0.9 nmol l^{-1}	[44]
Fish	Zn, Cd, Cu, Pb, and Hg	Zn, Cd, Cu, Pb, and Hg	Hanging mercury drop electrode	1–10 ng l^{-1},	[45]
Rice	Cd	DPASV	Au nanoparticle modified CPE	1.94 nmol l^{-1}	[46]
Chicken, duck, and turkey	Roxarsone (ROX)	DPV	SPCE modified with lanthanum molybdates	12.4 nmol l^{-1}	[47]
Apple, potato and lemon	Cd	SWV	PGE/bimetal oxide nanoparticles/graphene oxide	1.85 ng l^{-1}	[48]
Beans and corns	Pb, Cd, and Cu	LSASV	GCE	0.1 mol l^{-1}	[49]
Mushroom	Zn, Cu	DPASV	Amalgam electrode	10.3 mg kg^{-1}; 72.9 mg kg^{-1}	[50]
Canned tuna fish	Hg	DPV	GCE/Cu-MOF nanocubes	0.0633 nmol l^{-1}	[51]

LSASV: Linear Sweep Anodic Stripping Voltammetry, DPV: Differential Pulse Voltammetry, SWV: Square Wave Voltammetry, DPASV: Differential Pulse Anodic Stripping Voltammetry, GCE: Glassy Carbon Electrode, SPCE: Screen Printed Carbon Electrode, PGE: Pencil Graphite Electrode, MWCNT: Multiwalled Carbon Nanotubes, MOF: Metal Organic Framework.

Table 1.
Literature survey of metal analysis in food samples.

antioxidant activity using electrochemical sensors. The development of research in this area has been outlined with its initial stages in 1999, thereafter gaining momentum in 2010 and has remained so ever since. A total of 758 articles were published during this period. Electrochemical methods were used for the first time mainly for quantitative analysis, as well as other analytical approaches. Subsequently, CV was used to directly measure the electrochemical properties of various antioxidants and evaluate their antioxidant capacity. There were several advantages in this scenario when compared to the conventional DPPH assay [54]. The most direct application scenario for the evaluation of antioxidant capacity is in the food industry as it can have a direct impact on their price and nutritional value. One of the most widespread and produced antioxidants present commonly in food is vitamin C, which is electrochemically active and thus can be detected by electrochemical sensors. The development of an electrochemical sensor for Vitamin C brought a major breakthrough in the field of analytical chemistry [55]. Bounegru and Apetrei (2020) reported a nanomaterial based voltammetric sensor for the qualitative and quantitative determination of caffeic acid using CV (**Figure 1**). Carbon nanofibre (CNF) and MWCNT modified carbon based SPEs (C-SPE) were utilized to study the electrochemical behavior of caffeic acid in aqueous solution (pH 3.6). The LOD and LOQ values were seen to be in the range of 10^{-7} – 10^{-9} M (**Figure 1**) which indicates good sensitivity and in fact the

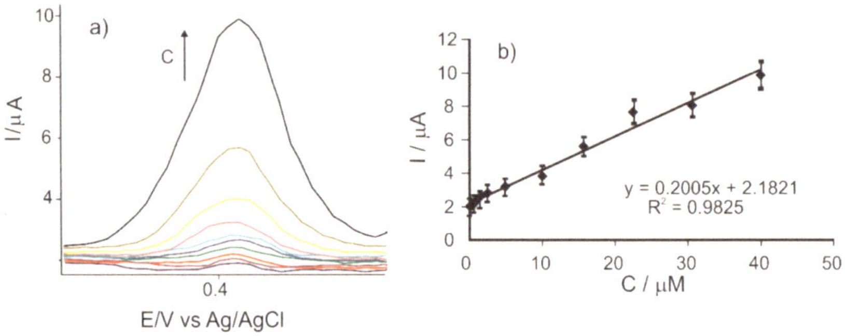

Figure 1.
(a) Zoomed-in view of the anodic peak zone of the CV registered with CNF/C-SPE immersed in caffeic acid solutions with the concentrations in the 0.1–40 µM range. (b) Linear dependence between the anodic peak current and the concentration of the caffeic acid solution [56].

electrochemical results were also compared to the spectrophotometric data. Also, among the two naomaterials used CNF based sensor proved to be better in terms of sensitivity and performance. The optimized sensor (CNF/C-SPE) was also tested for real samples (Active Detox (Herbagetica), DVR-Stem Glycemo (DVR Pharm) and green tea (Alevia) as depicted in **Figure 2** yielding satisfactory results [56].

Qin et al. (2020) reported a TiO_2/electro reduced graphene oxide (TiO_2/ErGO) nanoparticles based electrochemical sensor for the simultaneous analysis of ponceau 4R and tartrazine. The nanocomposite was prepared by ultrasonically dispersing TiO_2 nanoparticles in the grapheme oxide solution followed by an electro-reduction step. The TEM analysis (**Figure 3**) indicated the uniform distribution of the TiO_2 nanoparticles in ErGO nanoflakes confirming the nanocomposite formation which resulted in enhanced adsorptive stripping DPV current response as is expected because of the synergistic effect of the nanomaterials. The two colorants could be detected in a nanomolar range using the respective sensor with selectivity, sensitivity and stability. The sensor was also applied for real sample which was orange juice in this case [57].

2.2 Voltammetric sensing of drug & pharmaceuticals

As pharmaceutical firms evolve to keep pace with expanding population of humans globally, the global pharmaceutical industry reached $1.25 trillion in 2019 and is expected to reach $1.5 trillion by 2023. The implications of this data suggest that the pharmaceutical sales grew by 207.9 percent between 2005 and 2019. As a result, two main problems emerged: many drugs are misused/abused thus pressing the need for researching economical, portable, and effective sensors for monitoring drug overdose or biomedical monitoring; and secondly, tracking or monitoring of pharmaceutical or other contaminants in water sources to maintain human and ecological health is also necessity of today's world [58]. Based on the unique electrochemical interfaces between the nanomaterials and the analyte at hand, there are several approaches that may be more suitable for the detection of pharmaceuticals [59]. Voltammetry, potentiometry, amperometry, and electrochemical impedance spectroscopy (EIS) are examples of common electrochemical methods. Due to their higher sensitivity than CV, pulse methods such as normal pulse voltammetry (NPV), DPV and SWV are more commonly utilized for electrochemical detection. These pulse approaches

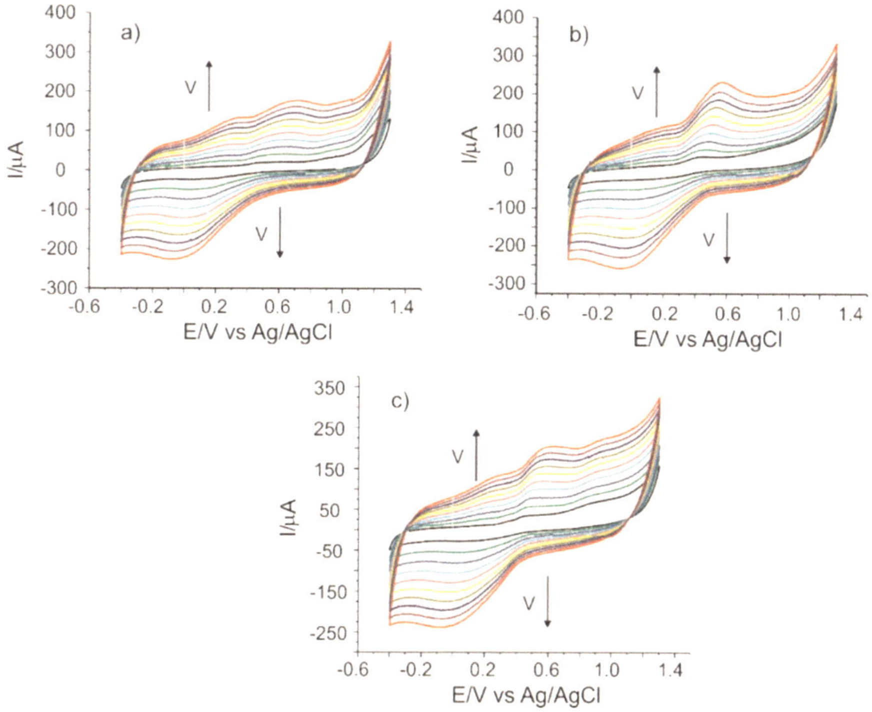

Figure 2.
CVs of CNF/C-SPE immersed in solutions of (a) active detox, (b) DVR-stem Glycemo and (c) green tea, recorded at scan rates between 0.1 and 1.0 $V{\cdot}s^{-1}$ [56].

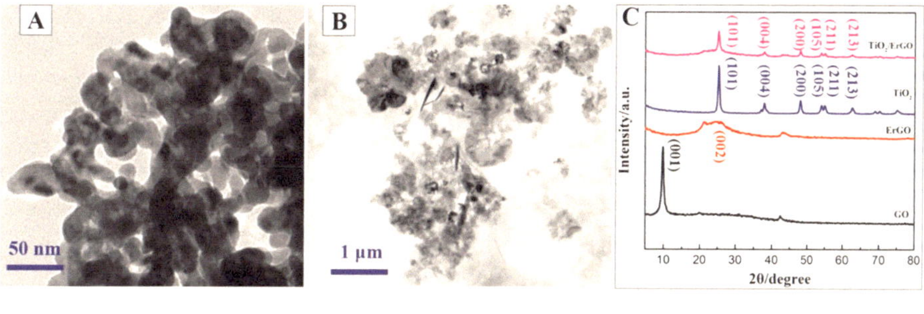

Figure 3.
TEM images of TiO_2 nanoparticles (A) and TiO_2/GO nanocomposites (B); XRD patterns of nanomaterials (C) [57].

significantly increase the signal of interest corresponding to the analytes compared to the capacitive current response [60].

Voltammetry and polarography are the most commonly used electroanalytical methods in pharmaceutical and biological analysis. In the 1930s and 1940s, the first examples of pharmaceutical analysis using polarographic techniques were documented. Most pharmaceutically active chemicals have been found to be electrochemically active [61]. Electrochemistry is a well-established and rapidly growing field with numerous potential applications in the pharmaceutical industry as is evident from **Table 2** [56]. The use of solid electrodes in the voltammetric determination

Sample	Target	Technique	Electrode	Drug	Linear Range	LOD	Reference
Non toxic Material	HCl	DPV	CPE	Dexamethosone	0.5–1 μg l^{-1}	4,10 μg l^{-1}	[62]
Non toxic Material	HCl	DPV	CPE	Prednisolone	0.5–1 μg l^{-1}	4,10 μg l^{-1}	[63]
Non toxic Material	HCl	DPV	CPE	Hydrocortisone	0.5–1 μg l^{-1}	4,10 μg l^{-1}	[64]
MWCNT	HCl	CV	CPE	Tinadozol	4.9 μg l^{-1}	8.3 μg l^{-1}	[65]
River Water	Buffer	CV	SPE	Ibufuran	1.6 μg l^{-1}	4.7 μg l^{-1}	[66]
River Water	Buffer	DPV	SPE	Flunitrazepam	2 μg l^{-1}	1 μg l^{-1}	[67]
Coca cola	Buffer	DPV	SPE	esomeprazole	3.5 μg l^{-1}	10 μg l^{-1}	[68]
Alco-pop	Buffer	DPV	SPE	gemifloxin	0.4 μg l^{-1}	2 μg l^{-1}	[69]

Table 2.
Literature survey of electrochemical analysis in drug & pharmaceuticals.

of drugs is gaining popularity due to its simple modification providing a large electrochemical active surface area for fast charge transfer kinetics.

Various modifiers were used to improve the recorded current intensity and the sensitivity of the electrode for the detection of pharmaceutical traces. Additionally, since the solid electrode is non-toxic, it can be used with minimal safety precautions. The phrase disposable electrode makes electrochemical drug determination easier than it was in the 1940s. This word is associated with solid electrodes, the most common home-made conventional solid electrodes, including carbon paste electrodes, screen-printed electrodes, boron-doped diamond electrodes, and pencil graphite electrodes. Both screen-printed and pencil graphite electrodes are single-use electrodes [70].

Mehmandoust et al. (2021) reported gold/silver core–shell nanoparticles (Au@Ag CSNPs) with conducting polymer poly(3,4-ethylenedioxythiophene) polystyrene-sulfonate (PEDOT: PSS) and functionalized multi-carbon nanotubes (F-MWCNTs) on glassy carbon electrodes to create a novel and sensitive voltammetric nanosensor for the first time to monitor trace levels of favipiravir [70]. Under optimized conditions and at a typical working potential of + 1.23 V (vs. Ag/AgCl), Au@Ag CSNPs/PEDOT: PSS/F-MWCNT/GCE revealed linear quantitative ranges from 0.005 to 0.009 and 0.009 to 1.95 μM with a detection limit of 0.046 nM (S/N = 3) with acceptable relative standard deviations (1.1–4.9 percent) for pharmaceuticals, urine samples, and human plasma without any sample pretreatment (1.12–4.93 percent) [71]. Amino acids, biologics, and antiviral drugs had little or no effect, and the sensing system exhibited exceptional reproducibility, repeatability, stability, and reusability. The results showed that this approach has potential applications in the diagnosis of FAV in clinical samples. The increased surface area and the synergistic interaction between the bimetallic core-shell nanoparticles and the entrapped carbon structures by the conducting polymer resulted in high sensing properties [72]. Ziyatdinova and Gimadutdinova (2021) developed a $CeO_2.Fe_2O_3$ nanoparticles based sensor for the estimation of lipoic acid which is extensively used in medicine as drug therapy. The SEM analysis (**Figure 4**) indicated the uniform distribution of nanoparticles on the electrode surface resulting in enhanced catalytic oxidation currents, reduced overpotential and improved electron transfer kinetics. The designed optimized sensor

(a) (b)

Figure 4.
Surface morphology of bare GCE (a) and $CeO_2.Fe_2O_3$ nanoparticles modified GCE [73].

(a) (b)

Figure 5.
SEM images of bare GCE (a) and SeO_2-CPB/GCE (b) [74].

was also successfully tested in pharmaceutical dosages for the lipoid acid [73]. Kavieva and Ziyatdinova (2022) subsequently reported a sensor based on SeO_2nanopartciles and surfactant (cetylpyridinium bromide; CPB) for the determination of indigo carmine which is a widely used colorant in pharmaceutical industry. The immobilization of SeO_2 nanoparticles on the electrode surface was confirmed with SEM analysis presented in **Figure 5**. The sensor was characterized using CV (**Figure 6**), chronoamperometry and electrochemical impedance spectroscopy (EIS) dictating reduction in charge transfer resistance at the fabricated sensor accompanied with roughly 4 times higher electroactive surface area [74].

2.3 Voltammetric biosensors

Since the first documented glucose biosensor more than half a century ago, biosensors have developed at an exponential rate. The use of biosensors in many fields such as diagnostics, environment, healthcare, and pharmaceuticals has resulted in the biosensor business growing into a multi-million dollar market that is expected to thrive in the coming years. In this massive biosensor sector, new advances in biosensors in terms of nanobiosensors should expand the range of such technologies in the market.

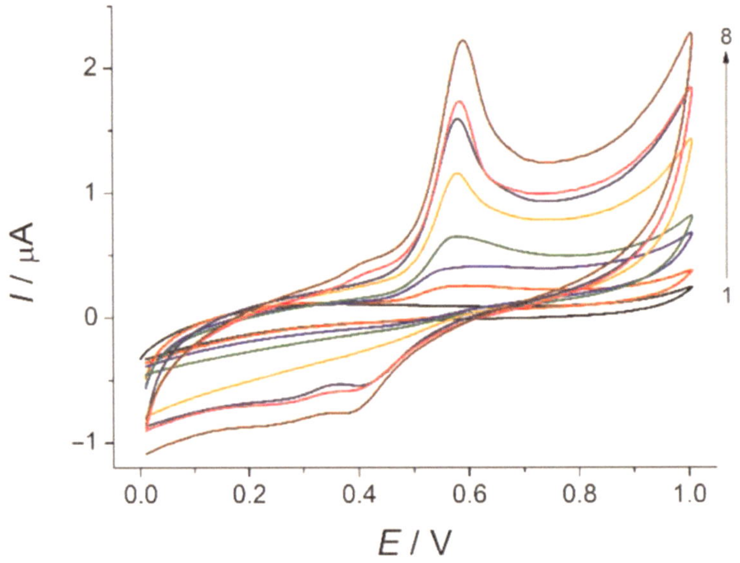

Figure 6.
CVs of 50 μM indigo carmine at SeO_2-CPB/GCE at various scan rates in pH 5 phosphate buffer [74].

Biosensors are generally based on the interaction of target species with the surface of a biomolecular receptor coupled to a transducer to generate a signal. A biosensor system can therefore be categorized based on how the detection was performed, such as electrochemical, mass, or optical. The method of detection can also be used to describe a nanobiosensor; however, the transducer used in such circumstances is usually from the field of nanotechnology, i.e. the use of nanoparticle or nanostructured interfaces. Optical nanoprobes detect various analytes using highly active metal nanoparticles such as gold or silver [75]. Although biosensors include well-established bioanalytical techniques, nanobiosensors have seen significant improvements in their sensitivity, resulting in a significant shift in the research trend from biosensors to the development of nanobiosensors. **Table 3** lists the range of electrochemical biosensors reported for analytes of importance in various sectors with their respective detection limit.

Kusior (2022) has reported a very interesting study for the determination of glucose without an enzyme receptor at copper oxide nanomaterial modified electrode wherein the facet exposition of Cu_2O nanoparticles has been linked to the current-potential profiles. The nanoparticles were synthesized using a wet chemical approach and different facets were exposed to current analysis for glucose determination (**Figures 7** and **8**). It is claimed that the electrochemical sensing or biosensing of Cu_2O nanoparticles depended on {100} and {111} facets, with former possessing more neutral states and the latter positive state. The study was performed by synthesizing particles in different sizes by addition of surfactants and the electrochemical performance was tested by CV and amperometry [76].

2.4 Voltammetric sensors for agriculture purposes

Contamination of pesticides and herbicides in soil, groundwater, rivers, lakes, storm water and air is a major problem. Pesticides are widely used in agriculture around the world and are a key tool for controlling weeds, insects and infections. They were defined as chemicals or mixtures designed to resist, eliminate, prevent or limit

Electrode	Modifier	Electrochemical Method	Analytes	Linear Range	Detection Limit	References
CPE	1,4-BBFT/IL	SWV	Isoproterenol	6.0×10^{-8}–7.0×10^{-4} M	12.0 nM	[77]
LIG	Salmonella typhimurium	SWV	Aptamer	10 CFU/mL	67–6.7×10^{5} CFU/mL	[78]
CPE	FC/CNT	DPV	N-acetylcysteine	1.0–400.0 μM	0.6 μM	[79]
SPE	S. pullorum & S. gallinarum	CV	Antibody	1.61×10^{1} CFU/mL	10^{1}–10^{9} CFU/mL	[78]
		DPV		3 CFU/mL	10–10^{7} CFU/mL	[80]
CPE	FC/MWCNT	DPV	Cysteamine	0.7–200 μM	0.3 μM	[81]
			folic acid	5.0–700 μM	2.0 μM	[82]
GME	S. typhimurium	CV	Au	2.4×10^{2} to 2.4×10^{7}	2.4×10^{2} cfu/mL	[83]
		DPV		3.1×10^{-5}–3.3×10^{-3} M	9.0×10^{-6} M	[84]
CPE	FC/CNT	DPV	Norepinephrine	0.47–500.0 μM	0.21 μM	[85]
CPE	2CBF/GO	SWV	Hydrochlorothiazide	5.0×10^{-8}–2.0×10^{-4} M	20.0 nM	[86]
CPE	FM/TiO_2 nanoparticle	DPV	Methyldopa	2.0×10^{-7}–1.0×10^{-4} M	8.0×10^{-8} M	[87]
CPE	FCD/CNT	DPV	Norepinephrine	0.03–500.0 μM	22.0 nM	[88]
CPE	2CBF/CNT	SWV	N-acetylcysteine	5.0×10^{-8}–4.0×10^{-4} M	2.6×10^{-8} M	[88]

Table 3.
Electrochemical biosensor developed for the detection of some analytes.

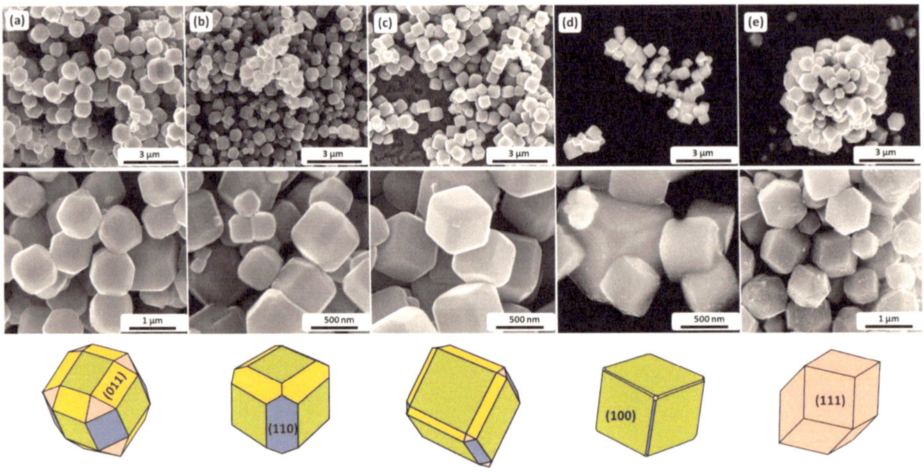

Figure 7.
SEM images of obtained polyhedral Cu_2O particles depending on surfactant used [76].

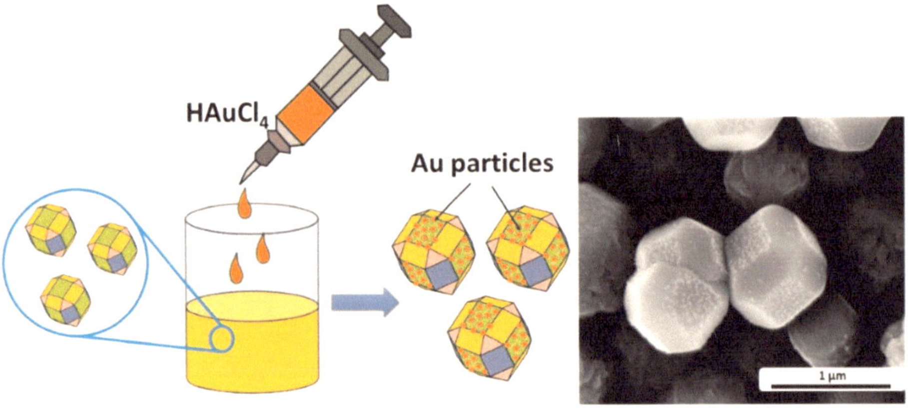

Figure 8.
Schematic representation of Au particle adsorption at the polyhedral surface with SEM image of modified Cu_2O grains [76].

the presence or effect of biological organisms capable of causing crop damage [89]. Since 1960, the increased use of pesticides has allowed farmers to significantly increase productivity while avoiding crop losses caused by pests [90]. Herbicides are pesticides that are used to kill or slow the growth of weeds. They were categorized based on their activity (contact or systemic), use (soil, pre-emergent or post-emergent) and mode of action on plant biochemical mechanisms [91, 92]. In addition, their target is classified as non-selective (kills all surrounding plants) or selective (attacks only weeds and leaves the crop alone) [93]. Glyphosate (N-(phosphonomethyl)glycine) was the most commonly used herbicide, followed by atrazine (2-chloro-4-ethylamino-6-isopropylamino-1,3,5-triazine) and 2,4D-dichlorophenoxyacetic acid) [94]. Priscila discussed the detection of herbicides (1990–2018) to obtain certain qualities and increase the detection limit, the matrix can be changed using metals, metal oxides, polymers, clay materials or micro or nanoparticles.

Sample	Transducers	Analyte	Biosensing Elements	Reference
Soil	Amperometric	2,4-D	Acetyl-Cholinesterase	[95]
Soil	Bioluminescense	Diuron, paraquat	Cyanobacterial	[96]
Soil	Potentiometric	Simazine	Peroxidase(Biocatalytic)	[97]
Soil	Impedance	2,4-D	Antibody	[99]
Soil	Potentiometric	Hydrazide	Acetyl-Cholinesterase	[101]
Drinking Water	Potentiometric	Isoproturon	Antibody	[100]
Waste Water	Potentiometric	Simazine	Peroxidase(Biocatalytic)	[102]
Orange Juice	Amperometric	Atrazine	Antibody	[98]

Table 4.
Sensors developed for the detection of some herbicides.

Electrochemical sensors are good candidates for environmental monitoring and have been successfully fabricated using various types of electrode materials. The applicability and advantages of voltammetric techniques over traditional methods have been well established. Surface modification of the electrode improves its properties, allowing small amounts of a specific analyte to be detected or quantified; for metals or metal oxide electrodes are now able to detect non-electroactive herbicides. Nanostructures have also been shown to have several advantages and are a promising field of study. There are a significant number of biosensor papers that provide exciting new potential to increase their performance for herbicide detection (**Table 4**). Most of the work presented used real samples such as tap and lake water, poisoned soil and food hazardous substances (fruits and vegetables), but there is still a need to incorporate these new devices into commercial analysis or industries.

2.5 Electrochemical energy sensors

Fuels are an essential part of today's civilization with significant economic and ecological importance. The current search for new sustainable alternative fuels to fossil fuels was based on minimizing the emission of pollutants into the atmosphere such as CO_2. Biofuels, as they were widely known, were commonly produced from vegetable oils and their composition can vary depending on their origin, resource types and their matrices. Metals and toxins in fuels, regardless of their origin, such as fossil or alternative, remain a problem that needs to be solved. A large number of metals can be used as catalysts in the oxidation processes of fuels when combined with gasoline, either by contamination or adulteration, and the results can damage automotive engine components [100]. Among many types of current fuels, it is customary to examine and assess the metal contamination in their composition, which exceeds the limitations of regulatory authorities to avoid economic and environmental losses associated with the concentration of specific metals in fuel matrices. There are several variables that affect the maximum amount of metal allowed in gasoline. It was determined by the regulatory body, the type of fuel and even the metal involved. Lead was not permitted in gasoline in Europe and manganese is limited to 2 mg/l [88]; however no metals are permitted in gasoline in the United States [101]. There was extensive literature on the evaluation of metals in fuels using a variety of approaches, most of which have a high material and resource analysis. Due to its mobility, high sensitivity and low cost compared to other approaches, electrochemical techniques offer a promising alternative route for these kinds of studies.

Dos Santos and Ferreira discussed in a 2020 publication the analysis of Pb^{2+} and Cu^{2+} in microemulsified biodiesel using a boron-doped diamond electrode [90]. This process has been used to produce biodiesel from corn oil, according to the researchers [102]. In comparison to other traditional detection methods, the results obtained were good. The analysis was performed despite the fact that these metals are not controlled in biodiesel. Ferreira and co-workers in 2021 used B-doped diamond electrodes in ethanol solution with ASV to measure Fe^{3+}, Cu^{2+}, Zn^{2+}, Pb^{2+} and Cd^{2+}. Combining interactions and interferences between species, was investigated [103]. A sample of fuel ethanol was tested in a hydroethanolic environment using this approach after the experimental parameters were tuned, with concentrations of investigated ions below the detection limits for the proposed method (**Table 5**).

2.6 Voltammetric environmental sensors

Some benefits are likely to emerge in the existing water monitoring environment when the mode of analysis moves based on sampling techniques to exclusively Conductivity-Temperature-Depth (CTD) probes have been used underwater since the 1980s, there has been a clear shift towards decentralized techniques approaches in a portfolio of known analytical techniques show clear promise for use in An evaluation of submersible devices in terms of their analytical properties, autonomy, miniaturization, and portability [111]. Plant modified carbon paste electrodes have been reported in literature recently for the estimation of heavy metals. One study was done in 2016 wherein bagasse (waste from sugarcane) was used as a biomaterial in a sensor to detect the hazardous metals Pb and Cd. An electrochemical approach was used to investigate the performance of the bagasse-based carbon paste sensor [112]. SEM, FTIR, CV and BET analysis were used to characterize the modifier's surface which indicated mesoporous pore distribution of bagasse and the specific surface area was calculated to be 89.3 m^2/g [100]. The accumulation was carried out in pH 6 acetate buffer and stripping analysis was done in HCl at a scan rate of 50 mV/s. The linearity was observed for lead and cadmium in the concentration ranges of 100–600 gL^{-1} and 500–1200 gL^{-1} with detection limits of 10.1 and 170.64 gL^{-1}, respectively for 10 mins of accumulated time.

Sample	Target	Technique	Electrode type	LOD	Reference
Biodiesel	Cu	DPASV	Hg film electrode	4.69 nmol-^{1}	[104]
Biodiesel	Pb	DPASV	Hg film electrode	2.91 nmol-^{1}	[105]
Bio-Ethanol fuel	Fe	LSSV	Nafion	2.4 μmol-^{1}	[106]
Bio-Ethanol fuel	Cu	DPSV	CNT	170 nmol l-^{1}	[107]
Biodiesel	Ca	SWV	GCE	1.6 nmol l^{-1}	[108]
Biodiesel	Tin	SWASV	Bi Film electrode	0.14 μmol l^{-1}	[109]
Bio-Ethanol fuel	Zn	SWASV	GDE	5 μg l^{-1}	[110]
Biofuel	Cu	SWASV	Vulcan functionalized	1.2 nmol l^{-1}	[111]

Table 5.
Electrochemical energy sensors for metal analysis.

Electrochemical sensors have proven ideal for this function, particularly in ion detection. Maria Cuartero reported (2021) an investigation of electrochemical sensors that showed promise for measuring ions in seawater such trace elements, nutrition, and carbon species The preceding five years were chosen as the major era for investigation, but older contributions to the area or goods introduced are included when important technical advancements are highlighted. There was a distinct absence of electrochemical sensors used in in-situ applications, which exacerbated when saltwater was considered: only a few examples have been demonstrated under such difficult conditions.

A glassy carbon electrode (GCE) modified by simultaneous electropolymerization of L-lysine (L-lys) and -cyclodextrin (-CD) film (P-CD-L-lys/GCE) was described as a novel electrochemical platform for the detection of pymetrozine. Using CV in 0.1 M H_2SO_4, the electrochemical activity of P-CD-L-lys/GCE towards pymetrozine was investigated [113]. The sensor's potential value was shown using LSV to estimate pymetrozine concentration analytically. The linear range is 4.0108 mol/L to 1.0106 mol/L. Pymetrozine's detection limit was determined to be 1.3108 mol/L (S/N = 3). We were able to acquire satisfactory pymetrozine results in real sample analysis using this approach. The usage of amino acid materials and cyclodextrin in environmental analysis was supported in this study. Chaiyo et al. (2020) reported a simple, low-cost, and highly sensitive voltammetric sensor based on a Nafion, ionic liquid, and graphene composite, modifying a screen-printed carbon electrode for the simultaneous determination of Zn^{2+}, Cd^{2+}, and Pb^{2+} in drinking water at the nanomolar level [114]. Del Valle and co-workers reported the immobilization of 4-carboxybenzyl-18-crown-6 and 4-carboxybenzo-15-crown-5 on monolayers of aryldiazonium salts anchored on the surface of graphite-epoxy composite electrodes for the simultaneous determination of Cd^{2+}, Pb^{2+} and Cu^{2+} in synthetic water samples at to the ppb level (ca. nanomolar) using differential pulse anodic stripping voltammetry (DPASV) [115]. Then, using a comparable electrode modification, Perez-Rafols et al. developed an electronic tongue for the detection of Cd^{2+}, Pb^{2+}, Tl^{+} and Bi^{3+} in synthetic water samples [101]. For the detection of trace metals, molecularly imprinted polymers [102–104] and metal-shaped nanoparticles [116, 117] have also been proposed. Voltammetric analysis of Pb^{2+} in enriched water has been reported using MoS_2/rGO flower composite with ultrathin nanosheets [118], Ni/NiO/MoO_3/chitosan 3D foam at the p-n junction interfacial barrier for micromolar level Cu^{2+} detection [119] and Mn-mediated MoS_2nanosheets have been reported recently as a new approach to Pb^{2+} sensing [120]. A VIP (Voltammetric In-Situ Profiler) for the detection of Cu^{2+}, Pb^{2+}, Cd^{2+} and Zn^{2+} using a Hg-based electrode [121] and a kayak equipped with Zn^{2+}voltammetric sensors based on liquid crystal polymer bismuth film [122] are the only two cases reported in the literature that demonstrated in-situ operation in seawater. Single and multiple analyte detection methods have been investigated in the last five years. Maria's (2021) report includes a number of electrochemical techniques for the detection of trace metals, although demonstrated applications in unspiked saline samples are quite limited (**Table 6**). For nitrogenous nutrients, two in-situ solutions were effectively used to produce NO_3 and NO_2 profiles in saltwater.

After seawater treatment, potentiometric detection of NO_3 and NO_2 at relatively low doses is possible utilizing multiple electrodes combined in a flow cell. Amperometric NO_x and NO_2 biosensing is possible using various bacterial chambers in an electrode configuration that can be deployed in any water system using an analyzer or submerged device allowing direct contact of the sensors with the water column at

Sensor	Analyte	Electrode	LOD	Reference
SWSV	Trace Metals Cu^{2+}	Hg	10^{-11} M	[123]
SWSV	Trace metal Pb^{2+}	Hg	10^{-11} M	[124]
SWSV	Trace metal Cd^{2+}	Hg	10^{-11} M	[125]
SWSV	Trace metal Zn^{2+}	Hg	10^{-11} M	[126]
SWSV	Trace metal Zn^{2+}	Bi	1 nM	[127]
ISE	Nutrients NO^{3-}	Iniline acid	1 μM	[128]
ISM	Nutrients NO^{2-}	Disalination	1 μM	[129]
ISM	TAN	GC/POT with nonactine	1 μM	[130]
SWV	Phosphate	Au,Mo-P	1 μM	[131]
ISM	Carbon species	GC/CNT/ISM with ionophores	μM, pH = 9	[132]
Acid base titration	Alkanility	Solid electrode	pH < 4	[133]
Amphoteric biosensor	NO_x^-	Bacteria chamber	1 μM	[134]

Table 6.
Electrochemical environmental sensors for metal analysis.

micromolar concentration with separate electrodes. There are no demonstrated in-situ uses of electrochemical devices for in-situ detection of ions in saltwater, according to studies. At the laboratory scale, new advancements are ongoing, with in-situ installation compatibility promised but rarely fulfilled. The few sensors that have been integrated in submersibles appear to be in the early stages of commercialization. As a result, meaningful data is only gathered through the inventors' measurements [135]. Electrochemical sensors have the potential to transform seawater analysis programs, but bringing this vision to fruition would need well-planned phases and targeted research.

2.7 Electrochemical sensing in forensics

Homicide cases usually involve weapons that release many compounds when fired. These compounds are identified in gunshot residue that may be found at a crime scene, on a suspects' clothing or other items, which then become useful in the investigation of a crime by providing evidence. This gunshot residue (GSR) consists of both inorganic and organic parts that need analysis for a successful investigation. In addition, elements such as antimony, lead, bismuth and copper present in GSR are hazardous to human health and the environment and therefore their determination is necessary [136]. Increasing security needs necessitate the deployment of field-deployable detectors capable of detecting GSR and nitro-aromatic explosive compounds in real time.

Erden et al. (2011) presented the measurement of antimony and lead in gunshot residues using differential pulsed cathodic adsorptive stripping (DPCAS) and square-wave cathodic adsorption stripping voltammetry (SWCASV) [137]. A hanging drop of mercury was used as a working electrode for GSR samples obtained during test firings at the Police Criminal Laboratory from the shooters' hands. The results demonstrated that both the DPCAdSV and SWCAdSV approaches could be used to successfully determine metals in GSR, both qualitatively and quantitatively. Vuki et al. (2012) then reported the simultaneous detection of Pb and Sb. along with other organic components of GSR at propellant on a glassy carbon electrode using CV and cyclic-SWV [138]. The thin film Hg GCE was used for Ba analysis because it requires detection of the presence of mercury [139].

Ceto et al. (2012) reported Zn, Pb, Cu, and nickel amalgams on bare and modified SPE by SWV in samples obtained straight from shooters at a nearby shooting range. The fascinating element is that they were able to validate several handling situations including secondary contact with the GSR, loading the handgun, and finally discharging. The subject that fired the firearm was identified other than the ability to distinguish the time of contact with the GSR with respect to different subjects. The use of microelectrodes for lead analysis in GSR was demonstrated by Salles et al. (2012) using the SWV technique [140]. Microelectrodes have substantial benefits for analysis, such as the capacity to deal with small-volume samples, downsizing of equipment, and the removal of mixing in the pre-concentration process. Their study suggested the possibility of linking lead originating from the GSR to discrimination between different types of weapons and ammunition. Wearable, completely stiff finger sensors for quick on-site voltammetric assessment of GSR and explosive surface residues were described by Bandodkar et al. (2013). To build new forensic fingers, they screen-printed the three-electrode arrangement onto a nitrile finger pad and overlay another finger pad with an ionogel electrolyte layer [141]. The novel integrated sampling/detection system uses "microparticle voltammetry" (VMP) to transmit minute quantities of surface-confined analytes directly onto a fingertip-based electrode array. Voltammetric measurements of sample residues are performed by bringing the working electrode into direct contact with a second finger bed coated in ionogel electrolyte (worn on the thumb), therefore completing the solid-state electrochemical cell [142]. The sampling and screening process took less than four minutes and results in GSR and explosives having distinct voltammetric signatures. The use of a solid, flexible ionogel electrolyte eliminates any liquid handling, reducing leakage, portability, and contamination concerns. The fingertip detection gadget demonstrated great specificity for detecting GSR and nitroaromatic chemical residues. It can tolerate constant mechanical stress without losing its attractiveness, a low-cost way for conducting on-site crime scene investigations in a range of forensic settings [143]. Hashim et al. 2016 reported gold modified screen printed electrode for Cu (II) analysis in GSR using CV (**Figure 9**). It was also compared with ICP-OES analysis with 94% accuracy [144].

The research area was long forgotten, but is now re-emerging as a result of the sensitivity and adaptability that this approach provides. Ott et al. published a paper in 2020 on the analysis of Pb, Sb, and Cu by SWASV employing bare SPE. Two samples were taken in this investigation, one from the hands of volunteers with no direct touch and the other from the hands of those who had recently discharged a weapon. They examined data from 395 genuine shooter samples and 350 background samples, making it the biggest GSR research yet done [145]. Using a simple, quick, and sensitive voltammetric assay, they were able to identify all metals as well as organic bullet residues at the same time. It was detailed how to make a biochar-modified carbon paste electrode. Oliveira et al. used DPAdSV in 2021 to assess lead ions in gunshot residue and hair coloring samples. GSR samples were acquired using an IMBEL 9 mm pistol and 100% cotton fabric, with no hair color specified. To eliminate interference from other metals contained in gunpowder components such as Sb (III), Cu (II), Cr (III), and Fe (II), this work employed the conventional addition technique to measure lead in the acquired samples [146].

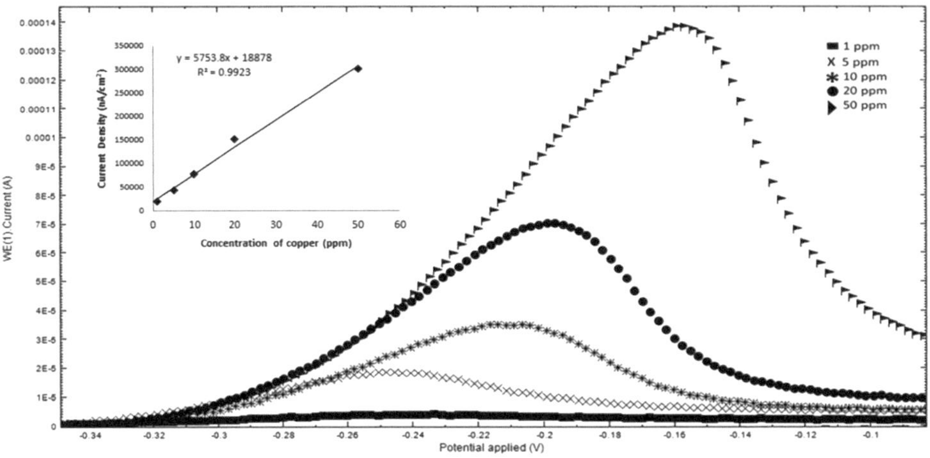

Figure 9.
CV at gold modified SPE for various Cu(II) concentration with inset showing calibration plot [144].

Bessa et al. (2021) presented a different approach to lead detection in *Lucilia cuprina*, a necrophagous insect that functioned as a GSR biomarker due to GSR consumption, As a result, an intriguing lead in forensic investigations is provided. Their survival rates were investigated, as well as the influence of lead on their embryonic growth [147]. As a working electrode, a bismuth film was deposited in an SPCE using the SAWASV process. The results showed that lead could be detected on larval samples even in the presence of interference such as Cu, Ba, and Cd. The effect of lead on mortality and larval development time was also successfully investigated [148, 149]. The revival of this area has opened up a whole new avenue for research leading to breakthrough in criminal investigations.

3. Conclusion

Voltammetry holds the key to future diagnostic tools for fast, real-time and sensitive analysis. It has found use in almost every spectrum of environment, medical care, agriculture, forensic, investigation, food and cosmetic industries for a range of analytes to be determined selectively or simultaneously depending on the requirement. The resurgence of the interest in electrochemical methods for the lucrative metal analysis required for various applications is derived from the advantages it caters to, like sensitivity and versatile application. Nanomaterials and polymers are the advanced materials being explored for their synergistic effect to modify electrode surfaces to overcome the challenges that stand in the way. The research is targeted to develop fast, stable, reproducible, miniaturized real-time sensors for use in various fields.

Acknowledgements

The authors are thankful to their respective organizations for the facilities provided to the authors to carry out their work. The authors would also like to acknowledge MDPI publisher for their open access journals which allowed us to reproduce content in form of figures.

Conflict of interest

The authors declare no conflict of interest.

Author details

Harsha Devnani[1*] and Chetna Sharma[2]

1 Department of Chemistry, School of Applied Sciences, Manav Rachna University, Faridabad, India

2 Department of Applied Chemistry, Aligarh Muslim University, Aligarh, India

*Address all correspondence to: harsha.devnani@gmail.com

References

[1] Majeed S, Naqvi STR, Najamul Haq M, Ashiq MN. Electroanalytical techniques in biosciences: Conductometry, coulometry, voltammetry, and electrochemical sensors. In: Egbuna C, Patrick-Iwanyanwu KC, Shah MA, Ifemeje JC, Rasul A, editors. Analytical Techniques in Biosciences. Elsevier, Academic Press; 2021. pp. 157-178. DOI: 10.1016/B978-0-12-822654-4.00004-X

[2] Power AC, Morrin A. Electroanalytical sensor technology. In: Khalid MAA, editor. Electrochemistry. London: Intech Open; 2013. DOI: 10.5772/51480

[3] Chen W, Cai S, Ren Q-Q, Wen W, Zhao Y-D. Recent advances in electrochemical sensing for hydrogen peroxide: A review. The Analyst. 2012; **137**:49-58. DOI: 10.1039/C1AN15738H

[4] Zhang X, Ju H, Wang J. Electrochemical Sensors, Biosensors and their Biomedical Applications. Elsevier, Academic press; 2008. pp. 307-581. DOI: 10.1016/B978-0-12-373738-0.X5001-6

[5] Stetter JR, Penrose WR, Yao S. Sensors, chemical sensors, electrochemical sensors, and ECS. Journal of The Electrochemical Society. 2003;**150**:S11. DOI: 10.1149/1.1539051

[6] Laina AL. Voltammetric Sensors for the Determination of Pharmaceuticals. Kochi: Cochin Institute of Science and Technology; 2013

[7] Devnani H, Ansari S, Satsangee SP, Jain R. ZrO_2-graphene-chitosan modified carbon paste for sensitive and selective determination of dopamine. Materials Today Chemistry. 2017;**4**:17-25. DOI: 10.1016/j.mtchem.2017.02.004

[8] Ansari S, Ansari MS, Devnani H, Satsangee SP, Jain R. CeO_2/g-C_3N_4nanocomposite: A perspective for electrochemical sensing of antidepressant drugs. Sensors and Actuators B: Chemical. 2018;**273**: 1226-1236. DOI: 10.1016/j.snb.2018.06.036

[9] Fojta M, Jelen F, Hevrek L, Palecek E. Electrochemical stripping techniques in analysis of nucleic acids and their constituents. Current Analytical Chemistry. 2008;**13**:250-262. DOI: 10.2174/157341108784911415

[10] Morrin A. Inkjet printed electrochemical sensors. In: Korvink JG, Smith PJ, Shin D-Y, editors. Inkjet Based Micromanufacturing. 1st ed. Weinheim: Wiley-VCH; 2012. DOI: 10.1002/9783527647101.index

[11] Newman JD, Turner APF. Home blood glucose biosensors: A commercial perspective. Biosensors Bioelectronics. 2005;**20**:2435-2453. DOI: 10.1016/j.bios.2004.11.012

[12] Jiang Z, Feng B, Xu J, Qing T, Zhang P, Qing Z. Graphene biosensors for bacterial and viral pathogens. Biosensor & Bioelectronics. 2020;**166**:112471. DOI: 10.1016/j.bios.2020.112471

[13] Vermisoglou E, Panacek D, Jayaramulu K, Pykal M, Frebort I, Kolar M, et al. Human virus detection with graphene-based materials. Biosensors & Bioelectronics. 2020;**166**: 112436. DOI: 10.1016/j.bios.2020.112436

[14] Kim K, Lee CH, Park CB. Chemical sensing platforms for detecting trace-level Alzheimer's core biomarkers. Chemical Society Reviews. 2020;**49**: 5446-5472. DOI: 10.1039/D0CS00107D

[15] Ligler FS, Gooding JJ. Lighting up biosensors: Now and the decade to come. Analytical Chemistry. 2019;**91**: 8732-8738. DOI: 10.1021/acs.analchem.9b00793

[16] Idilia A-CN, Ploense KL, Csordas AT, Kuwahara M, Kippin TE, et al. Seconds-resolved pharmacokinetic measurements of the chemotherapeutic irinotecan in situ in the living body. Chemical Science. 2019;**10**:8164-8170. DOI: 10.1039/C9SC01495K

[17] Li H, Dauphin-Ducharme P, Ortega G, Plaxco KW. Calibration-free electrochemical biosensors supporting accurate molecular measurements directly in undiluted whole blood. Journal of American Chemical Society. 2017;**139**:11207-11213. DOI: 10.1021/jacs.7b05412

[18] Quesada-González D, Merkoçi A. Nanomaterial-based devices for point-of-care diagnostic applications. Chemical Society Reviews. 2018;**47**:4697-4709. DOI: 10.1039/C7CS00837F

[19] Muniandy S, The SJ, Thong KL, Thiha A, Dinshaw IJ, Lai CW. Carbon nanomaterial-based electrochemical biosensors for foodborne bacterial detection. Critical Reviews in Analytical Chemistry. 2019;**49**:510-533. DOI: 10.1080/10408347.2018.1561243

[20] Curulli A. Nanomaterials in electrochemical sensing area: Application and challenges in food analysis. Molecules. 2020;**25**:5759. DOI: 10.3390/molecules25235759

[21] Patel BR, Nouroozifar M, Kerman K. Review-nanocomposite-based sensors for voltammetric detection of hazardous phenolic pollutants in water. Journal of The Electrochemical Society. 2020; **167**:037568. DOI: 10.1149/1945-7111/ab71fa

[22] Devnani H, Satsangee SP, Jain R. A novel graphene-chitosan-Bi2O3 nanocomposite modified sensor for sensitive and selective determination of a monoamine neurotransmitter epinephrine. Ionics. 2016;**22**:943-956. DOI: 10.1007/s11581-015-1620-y

[23] Silva LP, Silva TA, Moraes FC, Fatibello-Filho O. A voltammetric sensor based on a carbon black and chitosan-stabilised gold nanoparticle nanocomposite for ketoconazole determination. Analytical Methods. 2021;**13**:4495-4502. DOI: 10.1039/D1AY01321A

[24] Hassasi S, Hassaninejad-Darzi SK, Vahid A. Production of copper-graphenenanocomposite as a voltammetric sensor for anti-diabetic metformin using response surface methodology. Microchemical Journal. 2022;**172**:106877. DOI: 10.1016/j.microc.2021.106877

[25] Ismail NS, Hoa LQ, Huong VT, Inoue Y, Yoshikawa H, Saito M, et al. Electrochemiluminescence based enzymatic urea sensor using nanohybrid of isoluminol-gold nanoparticle-graphene oxide nanoribbons. Electroanalysis. 2017;**29**:938-943. DOI: 10.1002/elan.201600477

[26] Ponnaiah SK, Periakaruppan P, Vellaichamy B. New electrochemical sensor based on a silver-doped iron oxide nanocomposite coupled with polyaniline and its sensing application for picomolar-level detection of uric acid in human blood and urine samples. Journal of Physical Chemistry B. 2018;**122**:3037-3046. DOI: 10.1021/acs.jpcb.7b11504

[27] Ferrag C, Kerman K. Grand challenges in nanomaterial-based electrochemical sensors. Frontiers in Sensors. 2020;**29**:583822. DOI: 10.3389/fsens.2020.583822

[28] Alizadeh T, Atashi F, Ganjali MR. Molecularly imprinted polymer nanosphere/multi-walled carbon nanotube coated glassy carbon electrode as an ultra-sensitive voltammetric sensor for picomolar level determination of RDX. Talanta. 2019;**194**:415-421. DOI: 10.1016/j.talanta.2018.10.040

[29] Wu Y, Tilley RD, Gooding JJ. Challenges and solutions in developing ultrasensitive biosensors. Journal of American Chemical Society. 2019;**141**: 1162-1170. DOI: 10.1021/jacs.8b09397

[30] Gupta P, Tsai K, Ruhunage CK, Gupta VK, Rahm CE, Jiang D, et al. True picomolar neurotransmitter sensor based on open-ended carbon nanotubes. Analytical Chemistry. 2020;**92**: 8536-8545. DOI: 10.1021/acs.analchem.0c01363

[31] Tseng RC, Chen CC, Hsu SM, Chuang HS. Contact-lens biosensors. Sensors. 2018;**18**:2651. DOI: 10.3390/s18082651

[32] de Castro LF, de Freitas SV, Duarte LC, de Souza ACJ, Paixão TR, TomazelliColtro WK. Salivary diagnostics on paper microfluidic devices and their use as wearable sensors for glucose monitoring. Analytical Bioanalytical Chemistry. 2019;**411**: 4919-4928. DOI: 10.1007/s00216-019-01788-0

[33] Baghelani M, Abbasi Z, Daneshmand M, Light PE. Non-invasive continuous glucose monitoring system using a chipless printable sensor based on split ring microwave resonators. Science Reports. 2020;**10**:12980. DOI: 10.1038/s41598-020-69547-1

[34] Lichtenberg JY, Ling Y, Kim S. Non-specific adsorption reduction methods in biosensing. Sensors. 2019;**19**:2488. DOI: 10.3390/s19112488

[35] Li Y, Xu Y, Fleischer CC, Huang J, Lin R, Yang L, et al. Impact of anti-biofouling surface coatings on the properties of nanomaterials and their biomedical applications. Journal of Material Chemistry B. 2018;**6**:9-24. DOI: 10.1039/C7TB01695F

[36] Li T, Li Y, Zhang T. Materials, structures, and functions for flexible and stretchable biomimetic sensors. Accounts of Chemical Research. 2019;**52**: 288-296. DOI: 10.1021/acs.accounts.8b00497

[37] Meng LA, Turner PF, Mak WC. Soft and flexible material-based affinity sensors. Biotechnology Advances. 2020; **39**:107398. DOI: 10.1016/j.biotechadv.2019.05.004

[38] Malitesta C, Di Masi S, Mazzotta E. Electrochemical biosensors to biomimetic sensors based on molecularly imprinted polymers in environmental determination of heavy metals. Analytical chemistry. 2017;**10**:3389. DOI: 10.3389/fchem.2017.00047

[39] Mahale RS, Shashanka R, Vasanth RSV. Voltammetric determination of various food azo dyes using different modified carbon paste electrodes. Biointerface research in applied. Chemistry. 2022;**12**:4. DOI: DOI. 10.33263/BRIAC124.45574566

[40] Saidur MR, Aziz ARA, Basirun WJ. Recent advances in DNA based electrochemical biosensors for heavy metal ion detection. Review Biosensor and Bioelectronics. 2017;**90**:125-139. DOI: 10.1016/j.bios.2016.11.039

[41] Uloma A, Ihugba CO, Nwoko FR, Tony-Njoku AA, Ojiaku LI. Heavy metal determination and health risk assessment of oyster mushroom pleurotus tuber regium (Fr.) singer, collected from selected markets in Imo

State. NIGERIAN American. Journal of Environmental Protection. 2018;**6**:22-27. DOI: 10.12691/env-6-1-4

[42] Covaci E, Darvasi E, Ponta M. Simultaneous determination of Zn, Cd, Pb and Cu in mushroom by differential pulse anodic stripping voltammetry using thesis. Study University Babes-Bolyai University Of Cluj-Napoca. 2017; **3**:133-144

[43] Kupchyk O. Determining heavy metals in mushroom samples by stripping voltammetry. Food Science and Technology. 2018;**12**:2. DOI: 10.15673/fst.v12i2.939

[44] Raheem HAE, Hassan RYA, Khaled R, Farghali A, Sherbiny MIEL. New sensing platform of poly (ester-urethane) urea doped with gold nanoparticles for rapid detection of mercury ions in fish tissue. Royal Society of Chemistry. 2021;**11**:31845-31854. DOI: 10.1039/D1RA03693A

[45] Zhang L, Liao Q, Shao S, Zhang N, Shen Q, Liu C. Heavy metal pollution, fractionation, and potential ecological risks in sediments from Lake Chaohu (eastern China) and the surrounding Rivers. International journal Environmental Research Public Health. 2015;**12**:14115-14131. DOI: 10.3390/ijerph121114115O

[46] Veerakumar P, Sangili A, Manavalan S, Lin KC. Metal oxide-carbon nanocomposite-modified electrochemical sensors for toxic chemicals. ACS Applied Material & Interfaces. 2021;**2021**:173-212. DOI: 10.1016/B978-0-12-820727-7.00010-0

[47] Hashemi SE, Payehgadar M, Isaghi Z, Kargar H. Ultra trace level square wave anodic stripping voltammetric sensing of mercury(II) ions in environmental samples using a Schiff base-modified carbon paste electrode. International Journal of Environmental Analytical Chemistry. 2019;**99**:1148-1163. DOI: 10.1080/03067319.2019.1618455

[48] Singh S, Numan A, Zhan Y, Singh V, Van HT, Nam ND. A novel highly efficient and ultrasensitive electrochemical detection of toxic mercury (II) ions in canned tuna fish and tap water based on a copper metal-organic framework. Journal of Hazardous Materials. 2020;**399**:123042. DOI: 10.1016/j.jhazmat.2020.123042

[49] Gomes NO, Mendonca CD, Machado SAS, Olivera ONJR, Pereira PAR. Flexible and integrated dual carbon sensor for multiplexed detection of nonylphenol and paroxetine in tap water samples. Analytica Chimica Acta. 2021;**359**:188. DOI: 10.1007/s00604-021-05024-4

[50] Butmee P, Mala J, Damphathik C, Kunpatee K, Tumcharern G, Mehmeti KM, et al. A portable selective electrochemical sensor amplified with Fe_3O_4@Au-cysteamine-thymine acetic acid as a conductive mediator for determination of mercuric ion. Talanta. 2021;**221**:121669. DOI: 10.1016/2020.121669

[51] Kokulnathan T, Vishnuraj R, Wang TJ, Kumar EA, Pulithadahtil B. Heterostructured bismuth oxide/hexagonal-boron nitride nanocomposite: A disposable electrochemical sensor for detection of flutamide. Ecotoxicology & Environment Safety. 2021;**207**:111276. DOI: 10.1016/j.ecoenv.2020.111276

[52] Zabiha Pour T, Shahidi S-A, Karimi-Maleh H, Saraei G. A Voltammetric food analytical sensor for determining vanillin based on amplified $NiFe_2O_4$

nanoparticle/ionic liquid sensor. Journal of Food Measurement and Characterization. 2020;**14**:1039-1045. DOI: 10.1007/s11694-019-00353-8

[53] Amini K et al. Review—Recent advances in electrochemical sensor technologies for THC detection. Journal of Cannabis Research. 2022;**12**:4. DOI: 10.1186/s42238-022-00122-3

[54] Cantalapiedra A, Gismera MJ, Sevilla MT, Procopio JR. Sensitive and selective determination of phenolic compounds from aromatic plants using an electrochemical detection coupled with HPLC method. Phytochemistry Analytical. 2014;**25**:247-254. DOI: 10.1002/pca.2500

[55] Zhang P, Chun Z, Shao Q, Fu L, Luo Y, Gu D, et al. Evaluation of the phytochemicals and antioxidant activity of Lophatherum gracile Brongn based on chemical fingerprinting by HPLC with electrochemical detection. Journal of Seperation Science. 2021;**44**:3777-3788. DOI: 10.1002/jssc.202100318

[56] Bounegru AV, Apetrei C. Voltammetric sensors based on nanomaterials for detection of Caffeic acid in food supplements. Chem. 2020;**8**: 41. DOI: DOI.10.3390/chemosensors8020041

[57] Qin Z, Zang J, Liu Y, Wu J, Li G, Liu J, et al. A simple but efficient Voltammetric sensor for simultaneous detection of Tartrazine and Ponceau 4R based on TiO_2/electro-reduced graphene oxide nanocomposite. Chem. 2020;**8**:70. DOI: DOI.10.3390/chemosensors8030070

[58] Yang H, Zhou M, Yang W, Ren G, Ma L. Rolling-made gas diffusion electrode with carbon nanotube for electro-Fenton degradation of acetylsalicylic acid. Chemosphere. 2018; **206**:439-446. DOI: 10.1016/j.chemosphere.2018.05.027

[59] Rajendrachari S, BE K. Biosynthesis of silver nanoparticles using leaves of acacia Melanoxylon and their application as dopamine and hydrogen peroxide sensors. Physical chemistry. Research. 2020;**8**:25. DOI: 10.22036/PCR.2019.205211.1688

[60] Liu F, Zhao J, Wang S, Du P, Xing B. Effects of solution chemistry on adsorption of selected pharmaceuticals and personal care products (PPCPs) by Graphenes and carbon nanotubes. Environmental Science and Technology. 2014;**48**:13197-13206. DOI: 10.1021/es5034684

[61] Shashanka R, Swamy B. E K, simultaneous electro-generation and electro-deposition of copper oxide nanoparticles on glassy carbon electrode and its sensor application. SN Applied Sciences. 2020;**2**:956. DOI: 10.1007/s42452-020-2785-1

[62] Boxall ABA. The environmental side effects of medication: How are human and veterinary medicines in soils and water bodies affecting human and environmental health? European Molecular Biology Organisation. 2004;**5**:1110-1116. DOI: 10.1038/sj.embor.7400307

[63] Brody JG, Aschengrau A, McKelvey W, Swartz CH, Kennedy T, Ruthann AR. Breast cancer risk and drinking water contaminated by wastewater: A case control study. Environmental Health. 2006;**5**:28. DOI: 10.1186/1476-069X-5-28

[64] Sengar A, Vijayanandan A. Human health and ecological risk assessment of 98 pharmaceuticals and personal care products (ppcps) detected in Indian surface and wastewaters. Science Total Environmental. 2022;**807**:150677. DOI: 10.1016/j.scitotenv.2021.150677

[65] Kruc R, Dragon K, Gorski J. Migration of pharmaceuticals from the Warta River to the aquifer at a riverbank filtration site in Krajkowo (Poland). Water. 2019;**11**:1238. DOI: 10.3390/w11112238

[66] Hena S, Gutierrez L, Croué JP. Removal of pharmaceutical and personal care products (PPCPs) from wastewater using microalgae: A review. Journal of Hazardous Materials. 2021;**403**:124041. DOI: 10.1016/j.jhazmat.2020.124041

[67] Kuczynska J, Nieradko-Iwanicka B. The effect of ketoprofen lysine salt on mucosa of rat stomach after ethyl alcohol intoxication. Biomedicine & Pharmacotherapy. 2021;**141**:111938. DOI: 10.1016/j.biopha.2021.111938

[68] Cao F, Dong Q, Li C, Chen J, Ma X, Huang Y, et al. Electrochemical sensor for detecting pain reliever/fever reducer drug acetaminophen based on electrospun CeBiOx nanofibers modified screen-printed electrode. Sensors & Actuators B: Chemical. 2018;**256**: 143-150. DOI: /10.1016/2017.09.204

[69] Della Pelle F, Angelini C, Sergi M, Del Carlo M, Pepe A, Compagnone D. Nano carbon black-based screen-printed sensor for carbofuran, isoprocarb, carbaryl and fenobucarb detection: Application to grain samples. Talanta. 2018;**186**:389-396. DOI: 10.1016/j.talanta.2018.04.082

[70] Medsen KG, Skonberg C, Jurva U, Cornett C, Hansen SH, Johansen TN, et al. Bioactivation of diclofenac in vitro and In vivo: Correlation to electrochemical studies. Chemical Research in Toxicology. 2008;**21**: 1107-1119. DOI: 10.1021/tx700419d

[71] Asif AH, Wang S, Sun H. Hematite-based nanomaterials for photocatalytic degradation of pharmaceuticals and personal care products (PPCPs): A short review. Current Opinion in Green and Sustainable Chemistry. 2021;**28**:10044. DOI: 10.1016/j.cogsc.2021.100447

[72] Bhadra BN, Jhung SH. Adsorptive removal of pharmaceuticals and personal care products from water with functionalized metal-organic frameworks: Remarkable adsorbents with hydrogen-bonding abilities. Scientific Reports. 2016;**6**:34462. DOI: 10.1038/srep34462

[73] Kavieva L, Giyatdenova G. Voltammetric sensor based on SeO_2 nanoparticles and surfactants for indigo carmine determination. Sensors. 2022; **22**:3224. DOI: 10.3390/s22093224

[74] Giyatdenova G, Gimadutdenova L. Cerium(IV) and iron(III) oxides nanoparticles based Voltammetric sensor for the sensitive and selective determination of Lipoic acid. Sensors. 2021;**21**:7622-7639. DOI: 10.3390/s21227639

[75] Fan X, Gao J, Li W, Huang J, Yu G. Determination of 27 pharmaceuticals and personal care products (PPCPs) in water: The benefit of isotope dilution. Frontiers in Environmental Science Engineering. 2020;**14**:8. DOI: 10.1007/s11783-019-1187-3

[76] Kusior A. Voltammetric detection of glucose—The electrochemical behavior of the copper oxide materials with well-defined facets. Sensors. 2022;**22**:4783. DOI: 10.3390/s22134783

[77] Corcoran J, Tyler CR, Winter MJ. Pharmaceuticals in the aquatic environment: A critical review of the evidence for health effects in fish. Critical Reviews in Toxicology. 2010;**40**: 287-304. DOI: 10.3109/10408440903373590

[78] Moreno FA, Baeza JAM, Sanchez MG, Jimenez JG, Nguyen VAIC. Experimental validation of depth cameras for the parameterization of function of patient in clinical tests balance. Sensors. 2017;**17**:424. DOI: 10.3390/17020424

[79] Barbosa PFP, Viiera EG, Paim LL, Nakamura PR, Andrade RDA, Carmo DRD. Voltammetric techniques for pesticides and herbicides detection. International Journal of Electrochemical Science. 2019;**14**:3418-3433. DOI: 10.20964/2019.04.60

[80] Calixto CMF, Cervini P, Cavalheiro ETG. Determination of atenolol in environmental water samples and pharmaceutical formulations at a graphite-epoxy composite electrode. Materials Research. 2009;**10**:561-570. DOI: 10.1080/03067310903582358

[81] Fockmeniok SM etal. Sensitive carbon fibre microelectrode for the quantification of diuron in quality control of a commercialized formulation. International Journal of Analytical Chemistry. 2022;**22**:9994639. DOI: 10.1155/2022/9994639

[82] Prasertying P, Jantawong N, Sonsa-Ard T, Wongpakdee T, Khoonrueng N, Buking S, et al. Gold leaf electrochemical sensors: Applications and nanostructure modification. The Analyst. 2021;**146**: 1579-1589. DOI: 10.1039/D0AN02455D

[83] Ghaderpoori M, Jafari A, Nazari E, et al. Preparation and characterization of loaded paraquat- polymeric chitosan/ xantan/tripolyphosphatenano capsules and evaluation for controlled release. Journal of Environmental Health Science and Engineering. 2020;**18**:1057-1066. DOI: 10.1007/s40201-020-00527-3

[84] Simões FR, Mattoso LHC, VazCMP. Modified carbon paste-polyaniline electrodes for the electrochemical determination of the herbicide 2,4-D. Sensor Letters. 2004;**2**:221-225. DOI: 10.1166/sl.2004.049

[85] Mousty C. Sensors and biosensors based on clay-modified electrodes—New trends. Applied Clay Science. 2004;**27**: 159. DOI: 10.1016/j.clay.2004.06.005

[86] Qiu J, Villemure G. Anionic clay modified electrodes: Electrochemical activity of nickel(II) sites in layered double hydroxide films. Journal of Electroanalytical Chemistry. 1995;**395**: 159. DOI: /10.1016/0022-0728(95) 04070-5

[87] Paul PK, Hussain SA, Bhattacharjee D, Pal M. Preparation of polystyrene–clay nanocomposite by solution intercalation technique. Materials Research. 2013;**36**:361-366. DOI: 10.1007/s12034-013-0498-4

[88] Bouabi YE, Loudiki AK, Matrouf M, Akbour RA, Laghrib F, Farahi A, et al. Clay-based graphite sensor for electrochemical determination of p nitrophenol in water samples. Case Studies in Chemical and Environmental Engineering. 2022;**6**:100225. DOI: 10.1016/j.cscee.2022.100225

[89] Manisankar P, Selvanathan G, Vedhi C. Determination of pesticides using heteropolyacidmontmorillonite clay-modified electrodes with surfactant. Talanta. 2006;**68**:686-692. DOI: 10.1016/ j.talanta.2005.05.021

[90] Pifferi P, Soliveri G, et al. Electrochemical sensors cleaned by light: A proof of concept for on-site applications towards integrated monitoring systems. RSC Advances. 2015;**87**:1285. DOI: 10.1039/C5RA12219H

[91] Villulas HM, Costa FIM, Nascente PAP, Bulhoes LOS. Sol -gel

prepared Pt- modified oxide layers: Synthesis, characterization and electrolytic activity. Chemistry of Materials. 2006;**18**:5563-5570. DOI: 10.1021/cm0601178

[92] Mello RLS, Costa FIM, Villulas HM, Bulhoes LOS. Preparation and electrochemical characterization of Pt nanoparticles dispersed on niobium oxide. Analytical Chemistry. 2003;**28**:2. DOI: 10.1590/S0100-46702003000200009

[93] O'mahony AM, Samek IA, Sattayasamitsathit S, Wang J. Orthogonal identification of gunshot residue with complementary detection principles of voltammetry, scanning electron microscopy, and energy-dispersive X-ray spectroscopy: Sample, screen, and confirm. Analytical Chemistry. 2014;**86**:8031. DOI: 10.1021/ac5016112

[94] Bandodkar AJ, Mahony AMO, Ramírez J, Samek IA, Anderson SM, Windmiller JR, et al. Solid-state forensic finger sensor for integrated sampling and detection of gunshot residue and explosives: Towards 'lab-on-a-finger. The Analyst. 2013;**18**:138. DOI: 10.1039/c3an01179h

[95] Sassolas A, Prieto-Simón B, Marty JL. Biosensors for pesticides detection. American Journal of Analytical Chemistry. 2012;**3**:210. DOI: 10.4236/ajac.2012.33030

[96] Verma N, Bhardwaj A. Biosensor Technology for Pesticides. Applied Biochemistry and Biotechnology. 2003;**60**:29. DOI: 10.1007/s12010-015-1489-2

[97] Hayat A, Marty JL. Aptamer based electrochemical sensors for emerging environmental pollutants. Frontiers in Chemistry. 2014;**2**:41. DOI: 10.3389/fchem.2014.0004

[98] Rodriguez-Mozaz S, Maria J, Alda LDE, Barcelo D. Biosensors as useful tools for environmental analysis and monitoring. Analytical and Bioanalytical Chemistry. 2006;**386**:1025. DOI: 10.1007/s00216-006-0574-3

[99] Salgado AM, Silva LM, Melo AF. Biosensor for environmental applications. Environmental Biosensor. 2011;**386**:1025-1041. DOI: 10.5772/20154

[100] Webb S, Ternes T, Gibert M, Olejniczak K. Indirect human exposure to pharmaceuticals via drinking water. Toxicology Letters. 2003;**142**:157-167. DOI: 10.1016/s0378-4274(03)00071-7

[101] Navrátilová I, Skládal P. Theimmunosensors for measurement of 2,4-dichlorophenoxyacetic acid based on electrochemical impedance spectroscopy. Bioelectrochemistry. 2004;**42**:11-18. DOI: 10.1016/j.bioelechem.2003.10.004

[102] Maggioni S, Balaguer P, Chiozzotto C, Benfenati E. Screening of endocrine-disrupting phenols, herbicides, steroid estrogens, andestrogenicity in drinking water from the waterworks of 35 Italian cities and from PET-bottled mineral water. Environmental Science Pollution Research. 2013;**20**:1649-1660. DOI: 10.1007/s11356-012-1075-x

[103] Zwiener C. Occurrence and analysis of pharmaceuticals and their transformation products in drinking water treatment. Analyst Bioanalytical Chemistry. 2007;**387**:1159-1162. DOI: 10.1007/s00216-006-0818-2

[104] Tran-Minh C, Pandey PC, Kumaran. Studies on acetylcholine sensor and its analytical application based on the inhibition of cholinesterase. Biosensors and Bioelectronics. 1990;**5**:461. DOI: 10.1016/0956-5663(90)80035-C

[105] Martiniano LC et al. Direct simultaneous determination of Pb(II) and Cu(II) in biodiesel by anodic stripping voltammetry at a mercury-film electrode using microemulsions. Fuel. 2013;**103**:1164-1167. DOI: 10.1016/j.fuel.2012.07.002

[106] Almeida JMS et al. A simple electroanalytical procedure for the determination of calcium in biodiesel. Fuel. 2014;**26**:1794. DOI: 10.1016/j.fuel.2013.07.088

[107] Zhang Y, Zhuang HS. Amperometric immunosensor based on layer-by-layer assembly of thiourea and nano-gold particles on gold electrode for determination of naphthalene. Chinese Journal of Analytical Chemistry. 2010; **38**:153-157. DOI: 10.1016/S1872-2040(09)60021-9

[108] Ma R, Wang B, Lu S, Zhang Y, Yin L, Huang J, et al. Characterization of pharmaceutically active compounds in Dongting Lake, China: Occurrence, chiral profiling and environmental risk. Sci. Total Environ. 2016;**268**:557-558. DOI: 10.1016/j.scitotenv.2016.03.053

[109] Frena M, Campestrini I, OCDE B, Spinelli A. In situ bismuth-film electrode for square-wave anodic stripping voltammetric determination of tin in biodiesel. Electrochimica Acta. 2011;**56**: 4678. DOI: 10.1016/j.electacta.2011.02.111

[110] Oleveria TDC, Freiteies JM, Richter EM, Munoz RAA. A batch injection analysis system with square-wave voltammetric detection for fast and simultaneous determination of naphazoline and zinc. Talanta. 2016;**152**: 308-313. DOI: 10.1016/j.talanta.2016.02.031

[111] Niu X, Lan M, Zhao H, Chen C, Li Y, Zhu X. Review: Electrochemical stripping analysis of trace heavy metals using screen-printed electrodes. Analytical Letters. 2013;**46**:2479-2502. DOI: 10.1080/00032719.2013.805416

[112] Shashanka R, Jayprakash GK, Kumar M, Swamy BEK. Electrocatalytic determination of ascorbic acid using a green synthesized magnetite nano-flake modified carbon paste electrode by cyclic voltammetric method. Material Research Innovations. 2022;**26**:4. DOI: 10.1080/14328917.2021.1945795

[113] Erickson J et al. A simple and inexpensive electrochemical assay for the identification of nitrogen containing explosives in the field. Sensors. 2017;**17**: 1769. DOI: 10.3390/s17081769

[114] Tercier ML, Buffle J, Zirino A, Vitre RR. In situ voltammetric measurement of trace elements in lakes and oceans. Analytica Chimica Acta. 1990;**237**:429-437. DOI: 10.1016/s0003-2670(00)83947-1

[115] Tercier-Waeber ML, Buffle J, Graziottin F. A novel voltammetric in-situ profiling system for continuous real-time monitoring of trace elements in natural waters. Electroanalysis. 1998;**10**: 355-363. DOI: 10.1002/(SICI)1521-4109(199805)10:6<355::AID-ELAN355>3.0.CO;2-F

[116] Hughes G, Westmacott K, Honeychurch KC, Crew A, Pemberton RM, Hart JP. Recent advances in the fabrication and application of screen-printed electrochemical (bio) sensors based on carbon materials for biomedical, Agri-food and environmental analyses. Biosensors. 2016;**6**:50. DOI: 10.3390/bios6040050

[117] Tyszczuk-Rotko K, Szwagierek A. Green electrochemical sensor for caffeine determination in environmental

water samples: The bismuth film screen-printed carbon electrode. Journal of The Electrochemical Society. 2017;**164**:B342. DOI: 10.1149/2.0571707

[118] Wang N, Kanhere E, Kottapalli AGP, Miao JM, Triantafyllou MS. Flexible liquid crystal polymer-based electrochemical sensor for in-situ detection of zinc (II) in seawater. Microchimica Acta. 2017;**184**: 3007-3015. DOI: 10.1007/s00604-017-2280-6

[119] Farré MI, Pérez S, Kantiani L, Barceló D. Fate and toxicity of emerging pollutants, their metabolites and transformation products in the aquatic environment. Trends in Analytical Chemistry. 2008;**27**:991-1007. DOI: 10.1016/j.trac.2008.09.010

[120] Brillas E. A critical review on ibuprofen removal from synthetic waters, natural waters, and real wastewaters by advanced oxidation processes. Chemosphere. 2022;**286**: 131849. DOI: 10.1016/j.chemosphere.2021.131849

[121] Burns EE, Carter LJ, Snape J, Thomas-Oates J, Boxall ABA. Application of prioritisation approaches to optimize environmental monitoring and testing of pharmaceuticals. Journal of Toxicology Environment and Health B. 2018;**21**:115-141. DOI: 10.1080/10937404.2018.1465873

[122] Asif AH, Wang S, Sun H. Hematite-based nanomaterials for photocatalytic degradation of pharmaceuticals and personalcare products (PPCPs): A short review. Current Opinion in Green Sustainable Chemistry. 2021;**28**:100447. DOI: 10.1016/j.cogsc.2021.100447

[123] Nielsen M, Larsen LH, Jetten MSM, Revsbech NP. Bacterium-based NO_2^- biosensor for environmental applications. Applied Environmental Microbiology. 2004;**70**:6551-6558. DOI: 10.1128/AEM.70.11.6551-6558.2004

[124] Revsbech NP, Glud RN. Biosensor for laboratory and lander-based analysis of benthic nitrate plus nitrite distribution in marine environments. Limnology and Oceanographic Methods. 2009;**7**:761-770. DOI: 10.4319/lom.2009.7.761

[125] Bahnemann D. Photocatalytic water treatment: Solar energy applications. Journal Solar Energy. 2004;**77**:445-459. DOI: 10.1016/j.solener.2004.03.031

[126] Rodrigues PG, Gonçalves LM, Magalhães PJ, Pacheco JG, Rodrigues JA, Barros AA. Voltammetric analysis of metallothioneins and copper (II) in fish for water biomonitoring studies. Environmental Chemistry Letters. 2011; **9**:405. DOI: 10.1007/s10311-010-0293-z

[127] Castro SVF, Lima AP, Rocha RG, Cardoso RM, Montes RHO, Santana MHP, et al. Simultaneous determination of lead and antimony in gunshot residue using a 3D-printed platform working as sampler and sensor. Analytica Chimica Acta. 2020;**1130**:126. DOI: 10.1016/j.aca.2020.07.033

[128] Garcia-Segura S, Lanzarini-Lopes M, Hristovski K, Westerhoff P. Electrocatalytic reduction of nitrate: Fundamentals to full-scale water treatment applications. Applied Catalysis B: Environmental. 2018;**236**:546-568. DOI: 10.1016/j.apcatb.2018.05.041

[129] Legrand DC, Barus C, Garcon V. Square wave voltammetry measurements of low concentrations of nitrate using Au/AgNPs electrode in chloride solutions. Electroanalysis. 2017; **29**:2882-2288. DOI: 10.1002/elan.201700447

[130] Koyun O, Sahin Y. Voltammetric determination of nitrite with gold

nanoparticles/poly(methylene blue)-modified pencil graphite electrode: Application in food and water samples. Ionics. 2018;**24**:3187-3197. DOI: 10.1007/s11581-017-2429-7

[131] Sun C et al. Electrochemical sensor for nitrite using a glassy carbon electrode modified with gold-copper nanochainnetworks. Analytical Method. 2019;**19**:39. DOI: 10.1039/C9AY01544B

[132] Han YJ, Zhang R, Dong C, Cheng FQ, Guo YJ. Sensitive electrochemical sensor for nitrite ions based on rose-like AuNPs/MoS_2/graphene composite. Biosensors and Bioelectronics. 2019;**142**:111529. DOI: 10.1016/j.bios.2019.111529

[133] Hrastnik NI, Jovanovski V, Hocevar SB. In-situ prepared copper film electrode for adsorptive stripping voltammetric detection of trace Ni(II). Sensors & Actuators B: Chemical. 2020;**30**: 127637. DOI: 10.1016/j.snb.2019.127637

[134] Wang TT, Yue W. Carbon nanotubes heavy metal detection with stripping voltammetry: A review paper. Electroanalysis. 2017; **29**:2178-2189. DOI: 10.1002/elan.201700276

[135] Ustundag I, Erka AI, Koralay T, Kadioglu YK, Jeon S. Gold nanoparticles included graphene oxide modified electrode: Picomole detection of metal ions in seawater by stripping voltammetry. Journal of Analytical Chemistry. 2016;**71**:685-695. DOI: 10.1134/S1061934816070108

[136] Rajendrachari S, Ramakrishna D. 1 - functionalized nanomaterial-based electrochemical sensors: A sensitive sensor platform. Woodhead Publishing Series in Electronic and Optical Materials. 2022;**3**:25. DOI: 10.1016/B978-0-12-823788-5.00010-7

[137] Large RR, Halpin JA, Lounejeva E, Danyushevsky LV, Maslennikov VV, Gregory D, et al. Cycles of nutrient trace elements in the Phanerozoic Ocean. Gondwana Research. 2015;**28**:1282-1293. DOI: DOI.10.1016/j.gr.2015.06.004

[138] Daniel A, Laes-Huon A, Barus C, Beaton AD, Blandfort D, Guigues N, et al. Toward a harmonisation for using in situ nutrient sensors in the marine environment. Frontiers in Marine Science. 2020;**6**:773. DOI: 10.3389/fmars.2019.00773

[139] Han HT, Tao WY, Hu XP, Ding XY, Pan DW, Wang CC, et al. Needle shaped electrode for speciation analysis of copper in seawater. Electrochimica Acta. 2018;**289**:474-482. DOI: 10.1016/j.electacta.2018.08.097

[140] Zhang Y, Nie JT, Wei HY, Xu HT, Wang Q, Cong YQ, et al. Electrochemical detection of nitrite ions using Ag/Cu/MWNT nanoclusters electrodeposited on a glassy carbon electrode. Sensors & Actuators B: Chemical. 2018;**258**:1107-1116. DOI: 10.1016/j.snb.2017.12.001

[141] Trejos T, Vander C, Pyl M-HK, Arroyo AAL, LE. Fast identification of inorganic and organic gunshot residues by LIBS and electrochemical methods. Forensic. Chemistry. 2018;**8**:146. DOI: 10.1016/j.forc.2018.02.006

[142] Salles MO, Naozuka J, Bertotti M. A forensic study: Lead determination in gunshot residues. Microchemical Journal. 2012;**101**:49-53. DOI: 10.1016/j.microc.2011.10.004

[143] Somashekharappa KK, Rajendrachari S. Sustainable development information management of carbon nanomaterial-based sensors. Emerging Research Trends in Devices and Applications. 2022;**3**:12.

DOI: 10.1016/B978-0-323-91174-0.00001-9

[144] Hashim NA, Hashim M, Zain ZM, Jaafar MZ. Copper determination in gunshot residue by cyclic Voltammetric and inductive coupled plasma-optical emission spectroscopy. MATEC Web of Conferences. 2016;**59**:04005. DOI: 10.1051matec 59 conf/2016 04005

[145] Salles MO, Bertotti M, Paixão TRLC. Use of a gold microelectrode for discrimination of gunshot residues. Sensors and Actuators B: Chemical. 2012;**166-167**:848-852. DOI: 10.1016/j.snb.2012.02.097

[146] Spanu D, Binda G, Dossi C, Monticelli D. Biochar as an alternative sustainable platform for sensing applications: A review. Microchemical Journal. 2020;**159**:105506. DOI: 10.1016/j.microc.2020.105506

[147] Erden S, Durmus Z, Kiliç E. Simultaneous determination of antimony and Lead in gunshot residue by Cathodic adsorptive stripping Voltammetric methods. Electroanalysis. 2011;**23**:1967-1974. DOI: 10.1002/elan.201000612

[148] Vuki M, Shiu KK, Galik M, Mahony AM, Wang J. Simultaneous electrochemical measurement of metal and organic propellant constituents of gunshot residues. The Analyst. 2012;**137**: 3265-3270. DOI: 10.1039/c2an35379b

[149] Bessa BGDO, Silva-Neto HDEA, Coltro WKT, Rocha TL, Lopes WR. Lead toxicity in *Lucilia cuprina* and electrochemical analysis: a simple and low-cost alternative for forensic investigation. Analytical & Bioanalytical Chemistry. 2021;**413**:3201-3208. DOI: 10.1007/s00216-021-03257-z

Current and Emerging Challenges in the Diseases of Trees

Edited by Cristiano Bellé

Published in London, United Kingdom

Current and Emerging Challenges in the Diseases of Trees
http://dx.doi.org/10.5772/intechopen.102137
Edited by Cristiano Bellé

Contributors
Muqing Zhang, Palaniyandi Karuppaiya, Junyuan Huang, Frederick Kankam, Stephen Larbi-Koranteng, Joseph Adomako, Joseph Kwowura Kwodaga, Isaac Boatey Akpatsu, Yaw Danso, Elias Nortaa Kunedeb Sowley, Razak Terhem, Annya Ambrose, Jack Liam, Ángel Sol-Sánchez, Gloria Isela Hernández Melchor, Facundo Sánchez Gutiérrez, Cristiano Bellé, Rodrigo Ferraz F. Ramos, Lisiane Sobucki, Estéfany Pawlowski, Janaina S. Sarzi, Jessica E. Rabuske, Lucas G. Savian, Tiago E. Kaspary

First published in London, United Kingdom, 2023 by IntechOpen
IntechOpen is the global imprint of INTECHOPEN LIMITED, registered in England and Wales, registration number: 11086078, 5 Princes Gate Court, London, SW7 2QJ, United Kingdom

British Library Cataloguing-in-Publication Data
A catalogue record for this book is available from the British Library

Additional hard and PDF copies can be obtained from orders@intechopen.com

Current and Emerging Challenges in the Diseases of Trees
Edited by Cristiano Bellé
p. cm.
Print ISBN 978-1-80356-758-7
Online ISBN 978-1-80356-759-4
eBook (PDF) ISBN 978-1-80356-760-0

Meet the editor

Cristiano Bellé is a researcher in the field of plant nematology and biological control at Phytus Institute, Brazil. He is also a researcher at the Universidad Nacional de San Agustín de Arequipa (UNSA), Peru. He obtained a master's degree in Agronomy from the Federal University of Santa Maria, Brazil, in 2014, a doctorate degree in Crop Protection from the Federal University of Pelotas, Brazil, in 2018, and a post-doctorate degree in Soil Biology and Microbiology from the Federal University of Santa Maria in 2019. He has experience in the field of agronomy with an emphasis on plant pathology and plant nematology. His work is related to the etiology and diagnosis of plant diseases caused by nematodes and fungi, pathogenicity and host-parasite relationships of plant-parasitic nematode diseases, the interaction between plant-parasitic nematodes and soilborne fungi in plants, and biological, cultural, and chemical control of plant-parasitic nematodes.

Contents

Preface

This Edited Volume is a collection of reviewed and relevant research chapters, concerning the developments within the Current and Emerging Challenges in the Diseases of Trees field of study. The book includes scholarly contributions by various authors and is edited by a group of experts pertinent to Agricultural and Biological Sciences. Each contribution comes as a separate chapter complete in itself but directly related to the book's topics and objectives.

The Book includes the following chapters dealing with the topics: "Devious Phloem Intruder *Candidatus* Liberibacter Species Causing Huanglongbing: History, Symptoms, Mechanism, and Current Strategies", "Anthracnose Disease of Mango: Epidemiology, Impact and Management Options", "New and Emerging Disease Threats to Forest Plantations in Sarawak Borneo, Malaysia", "Main Pests and Diseases in Tropical Forest Species in Nursery" and "Perspective Chapter: Microorganisms and Their Relationship with Tree Health". The target audience comprises scholars and specialists in the field.

IntechOpen

Chapter 1

Devious Phloem Intruder *Candidatus* Liberibacter Species Causing Huanglongbing: History, Symptoms, Mechanism, and Current Strategies

Palaniyandi Karuppaiya, Junyuan Huang and Muqing Zhang

Abstract

Huanglongbing (HLB) or greening is a devastating phloem-intruding bacterial disease that generates various symptoms in leaves and fruits, threatening the global citrus industry. *Candidatus* Liberibacter asiaticus, *Candidatus* Liberibacter africanus, and *Candidatus* Liberibacter americanus are the causative agents of HLB in citrus-producing regions around many countries, and these proteobacteria are being vectorized by *Diaphorina citri* and *Triozaerytreae*. The lack of HLB-resistant citrus cultivars, the rapid spread of disease, and the fastidious nature of HLB-proteobacteria have made it difficult to mitigate HLB in the citrus field. There are numerous reports on the control of HLB disease using thermotherapy, chemotherapy, plant defense activators, brassinosteroids, and nanoemulsions. However, there is no evidence of such applicability of the methods mentioned above to complete the elimination or suppression of the pathogen to control HLB disease. We aim to provide an overall picture of HLB disease, its distribution, causal organism, pathogenic mechanism, and current and future strategies for combat against citrus Huanglongbing disease. This review may prompt the researchers toward an integrated and environmentally sustainable methodology for the mitigation/elimination of HLB pathogens.

Keywords: citrus, Huanglongbing, *Candidatus* Liberibacter, control strategy, psyllid

1. Introduction

Citrus fruits are the most predominantly produced fruits worldwide. The citrus species, Rutaceae family, is one of the major fruit crops in the world, which has provided an immune-enhancing source of vitamin C, nutrients, and medicinal value since ancient times [1]. Citrus crops are cultivated in more than 135 countries worldwide [2]. Worldwide citrus production is estimated at over 124.3 million tons annually [3]. Cultivated commercial citrus plants, consisting of rootstock and scion varieties, have a significant impact on scion growth, fruit quality, yield, and tolerance to biotic and abiotic stresses [4, 5]. Therefore, the selection of rootstock may make a significant contribution to the success or failure of the planting process [2]. However, various biotic and abiotic stresses impede citrus production worldwide, among which Huanglongbing is one of the significant pernicious diseases devastating the citriculture industry in the last few decades. Citriculture industries in Asia, Africa, and America have suffered massive economic losses due to the devastating Huanglongbing (HLB) malady [6].

Citrus HLB (Yellow dragon disease or citrus greening) is one of the highly ruinous diseases in citrus species caused by proteobacteria *Candidatus* Liberibacter species. The casual organisms of HLB have not been successfully cultured on axenic culture to date, and the prevalence of the HLB pathogen in citrus plants was evaluated using a diagnostic polymerase chain reaction (PCR) technique. *Diaphorina citri Kuwayama* (Asian citrus psyllid (ACP)) and *Triozaerytreae* (African citrus psyllid (AfCP)) transmit HLB disease from one citrus plant to another and also feed on many other species of the *Rutaceae* family [7]. The ACP resides in warm and humid zones and is most prevalent in Asia, the Indian subcontinent, Saudi Arabia, Reunion, and Mauritius. Now, ACP has also spread to South and Central America, such as Brazil, USA, and Mexico [8]. AfCP thrives in cold weather and is sensitive to the sweltering climate. AfCP resides in Africa, Cameroon, South Africa, Yemen, Madagascar, and Madeira Island [9]. HLB was first identified as a significant issue of unknown disease in citrus by farmers in southern China at the end of the nineteenth century [10]. HLB was first known a century ago as Citrus "Dieback" in India and "Yellow Dragon shoot Disease" in China, with a clear impact on citrus production in many countries, followed by South Africa, the Philippines, Indonesia, Thailand, Brazil, and the United States [11].

Citrus are susceptible to HLB, that is, nearly all commercial citrus and some citrus relatives. *Poncirus trifoliate* citrus, some *P. trifoliate* hybrids, and a few lemon varieties are considered more HLB tolerant [12]. The most efficient and sustainable strategy against citrus HLB is breeding resistant citrus cultivars. However, conventional citrus breeding is a long-term process that takes about 20 years to develop a new variety. Further, breeding efficiency is affected by gametophytic cross-incompatibility, heterozygosis, pollen-ovule sterility, apomixis, seedlessness, graft incompatibilities, polyembryony, and unstable characteristics [13]. Genetically engineered resistant citrus varieties are yet to be available for commercial cultivation due to the lack of acceptance of GMOs from farmers and consumers. It will, therefore, take many years to develop a promising resistant cultivar against HLB [14].

Many strategies to combat HLB were initiated, such as thermotherapy, antibiotics, plant defense initiators, pesticide, vector control management, chemotherapy, nanotechnology, and a transgenic approach [15, 16]. Beta-lactams, tetracyclines, and silver nanoparticles have obtained better results against HLB malady [17, 18]; however, the emergence of antibiotic resistance to microorganisms and indirect effects on human health and the environment is a significant and increasing risk that certainly restricts the use of antibiotics at the field level [18]. However, no effective strategies to eliminate or repress the HLB pathogens have been identified. This review attempts to provide an overall picture of HLB disease, distribution, casual organism and its pathogenic mechanism, and vector control management, and post the current and possible strategies to mitigate/combat HLB malady in the field.

2. Incidence, distribution, symptomology, and detection of citrus Huanglongbing

2.1 Incidence of Huanglongbing

HLB (also known as Yellow Dragon/shoot disease) was first identified as an unknown disease in citrus trees by citrus farmers in Guangdong Province, China, at the end of the nineteenth century [19], but studies suggest that HLB most likely originated in Taiwan in 1870 where it was known as Likubin ("Drooping disease") [20, 21]. Later, the HLB spread to other parts of China; by 1935, it had become a severe disease of citrus species [21]. Like HLB, dieback was first described in the central parts of India in the middle of the eighteenth century [22]. At that time, it might have been limited, but HLB was recorded in Assam in the eighteenth century and, by 1912, was a devastating disease in Bombay, India. However, the Citrus tristeza virus might cause this disease. Raychaudhuri [23] exhibited that dieback was the same as HLB. African greening disease was first identified in a sweet orange orchard in parts of South Africa in 1929 [24]. Outside of China, HLB was known as the "Greening" disease in South Africa, where extensive research was conducted in the 1950s. In Indonesia, the HLB disease was first noticed in the 1940s and is called the "citrus phloem degeneration" disease [25]. Reinking, in 1919, first described this disease in English as yellowing and leaf mottle of citrus noticed in China. According to International nomenclature rules, the name "Huanglongbing" was considered the official name by citrus pathologists at the 13th conference of the International Organization of Citrus Virologists in China [26]. "Huanglong" means yellowing of the shoot, as well as the yellow dragon (the symptom appears almost like a yellow dragon over the infected trees) and "bing," which means disease in Chinese [10]. Since the discovery of HLB, it has been named differently worldwide [27]. HLB was known outside under the name "citrus dieback" in India [23], "mottle leaf disease" in the Philippines [28], "vein phloem degeneration disease" in Indonesia [25], "yellow branch," "blotchy mottle," or "greening" disease in South Africa [29].

2.2 Geographical distribution

Globally, HLB has been considered one of the significant threats to citrus commercial and sustainable production. HLB was confirmed in citrus-producing regions of various countries, such as Nepal, Bangladesh, Thailand, Pakistan, Japan, Vietnam, Cambodia, Laos, Malaysia, Central African Republic, Comoros, Ethiopia, Kong Hong, Kenya, Madagascar, Malawi, Mauritius, Saudi Arabia, Reunion, Rwanda, Yemen, Zimbabwe, Somalia, Tanzania, Swaziland, and various region of United States of America including California, Florida [7, 27]. HLB has been reported in 24 countries and territories in East, South, Southwest Asia, East, and South Africa. Since then, it has been widely spread in other Asian, American, and African countries [27].

2.3 Symptomology

HLB symptoms are more evident in cold weather conditions than in hot seasons [30]. It is difficult to specify the period between when the citrus tree is affected by HLB and the onset of disease symptoms. It will exhibit in different parts of the plants or only in infected sectors when it eventually manifests symptoms. It is, therefore, difficult to diagnose and control at the early stage of HLB disease [31]. The HLB-infected tree exhibits symptoms in various parts of the plant depending on the stage of infection. If infection occurs soon after propagation, the entire tree gets affected and turns yellow all over the canopy, which leads to a decline irrevocably. Both the symptoms and the causative organisms were restricted to the infected sector in the event of later infection [27]. Only the infected sector will exhibit symptoms in the case of citrus trees affected by HLB, while the remaining parts will show normal growth and good-quality fruits. The symptoms observed on the HLB-affected tree include a heavy drop in the leaf and out-of-season flushing and blooming. Chronically, HLB-affected trees displayed stunting growth, twig dieback, sparse yellow foliage, or severe fruit drop [24]. The initial stage of HLB is vein yellowing [32], and the secondary level includes (infected leaves) small, upright with various chlorotic patterns similar to that caused by nutrient deficiency, such as zinc, sulfur, iron, boron, manganese, and calcium [33, 34]. In severe cases, the leaves were utterly void of chlorophyll, except for rounded green spots located on the leaves at random places [24]. The most accurate diagnostic symptom for HLB is that the infected fruits are small, lopsided, and taste bitter and salty. HLB-affected trees with premature shedding of green fruit drops while remaining on the tree, in which fruits with yellow halo-like lesions were staying green on the shaded side, hence the name "greening" [7, 34]. Root systems are developed in severely infected trees that exhibit poorly formed roots with few fibrous roots due to undernourishment [24] and repression of new root growth and rootlets decay [10].

HLB disease is challenging to diagnose based on symptoms, particularly during the early stages of the disease. Numerous symptoms of HLB might occur, and citrus trees are often caused by other diseases or nutrient deficiencies that may lead to similar symptoms [11, 30, 35]. Symptoms could be aggravated by other pathogens being coinfected. Several reports from Asian countries postulated that HLB-affected citrus trees are commonly coinfected with the *Citrus tristeza virus* (CTV) [7]. Interestingly, some CTV isolates protect trees against HLB infection [36]. Blotchy mottle leaf is a principal diagnosis of HLB that could be misinterpreted with other diseases, such as stubborn citrus disease caused by *Spiroplasma citri*, a severe infection of CTV phytophthora root rot, zinc deficiency, and waterlogging. Furthermore, it can also be confused with symptoms of leaf-related stress and mineral deficiency [37]. Early stages of citrus blight are also associated with the symptoms of zinc deficiency [38]. For these confusing symptoms of the disease, an unequivocal diagnosis technique is needed for HLB disease.

2.4 Method of HLB detection

Early identification and isolation of *Canditatus* Liberibacter species-infected trees are effective management approaches used to limit the spread of HLB from invading HLB disease-free citrus orchards in local and international trade [39]. Visual examination is one of the most commonly employed approaches for detecting citrus HLB disease. Traditionally, early detection of HLB disease relied primarily on various symptoms in the field, such as blotchy mottle leaf, yellow shoot, aborted seed, and lopsided fruit with green color remaining at the stylar end [40]. Nevertheless, this approach is highly affected by subjective interpretation, diagnostic errors can be higher than 30%, and other biotic and abiotic stress-related problems may worsen diagnosis. HLB symptoms might be confused with diseases such as *Citrus Tristeza*

Closterovirus, *Phytophthora* infection, citrus blight, and specific nutrient deficiencies [41]. Thus, the availability of advanced technologies that enable early and rapid detection of HLB pathogens is crucial [42]. Currently used methods for the diagnosis and confirmation of HLB disease include serology, enzymatic assay, enzyme-linked immunosorbent assays (ELISA), transmission electron microscopy, DNA probes, conventional polymerase chain reaction (PCR), quantitative PCR (qPCR), Fourier transform infrared spectroscopy (FTIR), and mid-infrared spectroscopy. Pereira et al. [43] developed a method for early diagnosis using X-ray fluorescence. The laser-induced breakdown spectroscopy (LIBS) combined with chemometric strategies is used to predict the condition of orchard plants infected with *Canditatus* Liberibacter species successfully. However, these methods did not provide early diagnosis except for the LIBS method. Recently, Tran et al. [44] reported a sensitive and selective label-free biosensor that combines the physical and chemical advantages of carbon nanomaterials such as single-walled carbon nanotubes (SWNTs) in a field-effect transistor (FET)/chemiresistor architecture with selective antibodies against Sec-delivered effector 1 (SDE1), a secreted protein biomarker, for the detection of HLB. Detailed HLB detection techniques have recently been reviewed [42, 45].

3. Causal agents of citrus Huanglongbing

The bacterium associated with citrus HLB was *Candidatus* Liberibacter species, which belongs to the alpha-proteobacteria determined by the 16 s ribosomal DNA sequences and the operon [9]. *Proteobacteria* associated with HLB disease in citrus are successively referred to as *Candidatus* Liberibacter asiaticus (Las) found in the majority of HLB-affected countries, *Candidatus* Liberibacter africanus (Laf) limited to African countries, and *Candidatus* Liberibacter americanus (Lam) limited to America [15]. How *Candidatus* Liberibacter bacterium established its association with citrus species remains unclear.

Scientific classification of *Candidatus* Liberibacter.
Kingdom: Bacteria.
Phylum: Proteobacteria.
Class: Alphaproteobacteria.
Order: Rhizobiales.
Family: Rhizobiaceae.
Genus: *Candidatus* Liberibacter.

3.1 *In vitro* culture of *Candidatus* Liberibacter species associated with HLB

The isolation of *Candidatus* Liberibacter species, causing HLB in an artificial culture medium, was a primary target for many researchers. Davis et al. [46] attempted to isolated *Candidatus* Liberibacter asiaticus in a culture medium from young angular green shoots from HLB-affected trees. A growth film appeared on the bottom of the tube containing broth AD medium. After single-colony isolation, Las and the actinobacteria closely related to *Propionibacterium acnes* remained together. Thus, Las was not isolated in axenic culture. Moreover, actinobacteria are prevalent residents of citrus and psyllids, whether Las is present. Sechler et al. [47] successfully cultivated a single colony of all three *Candidatus* Liberibacter species from HLB-affected leaf midveins and petiole sap in a new medium designated Liber A. The isolated cells were ovoid to rod-shaped, 0.3 to 0.4 by 0.5 to 2.0 μm, often with fimbriae-like appendages. They isolated two Las and one Lam strains from non-inoculated tissues of inoculated trees and seedlings 9 and 2 months later.

3.2 The pathogenic mechanism of *Candidatus* Liberibacter

Candidatus Liberibacter species are gram-negative, phloem-restricted bacterium associated with the pernicious disease of citrus HLB. Although *Candidatus* Liberibacters have been cultivated in artificial media, traditional molecular and genetic analyses have been difficult to perform owing to declining viability in culture [46, 48]. This difficulty has significantly limited efforts to comprehend the mechanisms of Liberibacter virulence. To date, most insights into the mechanisms of Liberibacter pathogenesis have been acquired through genomic analyses of Liberibacter sequences, host plant transcriptomic, proteomic, and metabolomic data associated with Liberibacter infection, and studies involving surrogates such as *Sinorhizobium* and *E. coli*, and expression in planta [15]. Evidence suggests that Liberibacters species associated with HLB live solely within the phloem tissues of host citrus plants [15]. Las bacterium resides inside a sieve tube and companion cells [47, 49]. The relatively consistent symptomology among various symptom-expressing hosts is one of the hallmarks of diseases caused by Liberibacter species [15].

3.2.1 Liberibacter secretion system and effector protein

The secretome of a pathogenic bacterium represents an array of molecules that play offensive roles during colonization, among which effectors are an important class of proteins capable of suppressing defense and/or manipulating host physiology [50, 51]. Interestingly, Las contain type I secretion systems (T1SSs) and a complete general secretory pathway (Sec), but lack other secretion systems (T2SS and T3SS) [15, 52], which play a significant role in extracellular pathogenic attacks on plant and animal host [16].

Liberibacter genome analyses found a complete T1SSs system in Las and Laf, but not in Lam [15]. Genes encode for serralysin and hemolysin; a T1SSs effector protein has been identified in Las and Laf genomes [53]. Serralysin is a metalloprotease secreted by gram-negative bacteria to inactivate peptides and antimicrobial proteins produced by the host plant. Las bacterium might use serralysin to degrade antimicrobial proteins in the host as its defense mechanism. This degraded protein is used for growth and metabolism by the Las bacterium as a carbon and nitrogen nutrient [16]. On the other hand, the hemolysin gene has been identified in all sequenced Liberibacters, which play an essential role in bacteria survival in the host plant. Las-produced hemolysin triggers ion leakage and water molecules from the host cell that lead to host cell apoptosis [16, 54].

The Secretary pathway (Sec) or Sec-translocon facilitates these effector proteins' transports outside the cytoplasm membrane vital for bacterial viability. The Sec machinery also secretes essential virulence factors in some plant-pathogenic bacteria [15]. *Candidatus* Liberibacter species have a general secretory pathway, which may lead to the secretion of effector proteins [55]. Since *Candidatus* Liberibacter species are phloem-resided bacteria, there is an inference that the bacteria secrete effector proteins directly into the cytoplasm of the host cells and modulate their physiology [56]. The effector protein CLIBASIA_05315 was located in transgenic citrus chloroplasts, resulting in leaf chlorosis and plant growth retardation [57]. Several research groups are currently focusing on identifying and characterizing the effector proteins of *Candidatus* Liberibacter species, and it is expected that we will have an improved view of this pathogenic mechanism of bacteria in a few years.

3.2.2 Lipopolysaccharides

Lipopolysaccharides (LPS), also known as endotoxin, are critical components derived from the outer membranes of gram-negative bacteria consisting of lipid A, an oligosaccharide core, and an O-antigen. LPSs are involved in outer membrane functions that are crucial for bacterial growth, survival against antimicrobial chemicals, and virulence, particularly within a host-parasite interaction. Lipid A is highly conserved, then the oligosaccharide core and O-antigen [15, 16]. LPSs are classical activators of defense responses in plants during plant-pathogen interaction [58]. Las bacteria use gene encoding active salicylate hydroxylase (SahA) to degrade salicylic acid (SA) and suppress plant defense mechanism. Intriguingly, the *SahA* gene is highly expressed *in planta*, while it is not expressed in psyllid vectors [56]. Las impedes SA-mediated defense responses in the phloem using its SA hydroxylase and maintains significant bacterial titer in citrus HLB disease progression over several years before the tree irrevocably declines. It is yet to be determined whether LPSs of Liberibacter cause callose accumulation in the phloem.

3.2.3 Flagella

The bacterial flagellum organelle, an intricate multiprotein essential for its rotational propulsion, promotes host colonization through adherence and induces plant immune modulation [15]. Las flagella have been reported to trigger host plant defense *in planta* as a pathogen-associated molecular pattern (PAMP) [59]. Microscopic studies found that flagella have not been observed in the *Candidatus* Liberibacter species that reside in the phloem in HLB-infected samples [11]. Despite the small size of the genome, genes associated with flagella biosynthesis have been identified in the sequenced Liberibacter genome [15, 16, 51, 52]. The genes *fliF*, *flgI*, and *flgD* expressed in flagellar assembly and the *motB* gene associated with the motor function were overexpressed *in planta*.

The *flbT*, an essential flagellin regulatory protein that acts as a regulatory checkpoint for flagellin gene expression, is found in the Las bacterial genome, whereas it is not in the Lam genome. The absence of *flbT* in the Lam genome results in no PAMP activation *in planta* [60]. Conversely, *flgL*, *flgK*, and *fliE* were overexpressed in psyllid [61].

3.2.4 Prophages

Several pathogenic bacteria harbor prophages or phage remnants integrated into their genome, encoding lysogenic genes that are proven or suspected virulence factors [59]. Las- and Lam-sequenced genome contains two potential prophages, Type 1 represents prophage SC1, and Type 2 represents prophage SC2. SC1 involved in the lytic cycle of forming phage particles. SC2 was implicated in the lysogenic conversion of Las pathogenesis [60, 62]. Type 3 prophage (P-JXGC-3) was identified in Las samples collected from Southern China. This prophage carries another bacterial defense system, such as a restriction-modification system (RM system) [63]. This RM system is fortified with endonucleases, which cleaves invading DNA that protects host DNA by altering specific sequences [64]. Type 1 and Type 2 prophages were not detected in the Las strain from Southern China. It is not clear whether these strains contain prophages or have unknown prophages. There are no comprehensive studies to describe the Las prophage repertoire [65]. Among strains observed in an extensive survey of Las isolates in China, it was typical for Las to have a single prophage, with Guangdong isolates harboring mainly the type 2 prophage, whereas isolates from Yunnan are dominated by the type 1 prophage [65]. The Las strain genome from Japan does not contain prophages [56]. Among the Las whole-genome sequences recently reported from different geographic areas around the globe, eight Las genomes contain extensive prophage sequences [63]. A survey of prophage prevalence in southern China revealed active prophage-phage interactions in the Las bacterial strains [63]. The exact function of the RM system has yet to be experimentally determined in Type 3 prophages. However, the lack of a prophage in many Las strains does not relate to the lack of HLB symptoms because Ishi-1 and the Guangdong isolates, which do not contain any prophages, induce similar HLB symptoms as isolates containing prophages [54, 65]. Overall, this evidence suggests that prophages contribute to bacterial virulence but are not required for Las pathogenicity.

3.3 Phloem dysfunction of HLB-affected citrus

Las bacteria reside within phloem and colonize sieve tubes [15, 16, 66]. Phloem dysfunction is a primary modification due to hyperactive differentiation of vascular cambium and hypertrophy of parenchyma cells surrounding the necrotic phloem pocket that may determine the development of HLB symptoms [32, 67]. HLB-associated Liberibacter secretes virulence factor and Sec-dependent effectors (SDEs) into phloem that stimulates HLB symptoms by interfering with either phloem or companion cell protein and genes of the host [15]. The secreted SDEs and virulence factors may interact with plastids, mitochondria, vacuole, and endoplasmic reticulum in the host phloem and target host genes and proteins to promote pathogen growth and disease development and suppress host immune responses [15]. Eventually, it leads to phloem malfunction in the host plant due to the Liberibacter virulence factors and SDE effects on sieve tubes and companion cells, which provide protein and transcripts to the sieve elements. Necrotic phloem was found in the HLB-infected plants due to starch (**Figure 1**) and callose deposition [32]. Callose accumulation was observed in sieve plates of Las-infected citrus [67]. Phloem dysfunction is generally associated with phloem sieve elements plugged with extensive deposition of callose and phloem protein 2 [67, 68], followed by phloem cell wall distortion and sieve element collapse [69]. Subsequently, photoassimilate transport was significantly blocked due to necrotic phloem [15, 16, 66, 68], which might result in substantial quantities of starch particles in almost all living cells of aerial parts, including phloem parenchyma and the sieve tube elements [32, 70]. The excessive accumulation of starch and zinc deficiency in chloroplast disrupts the thylakoid resulting in nonuniform loss of chlorophyll that triggers noticeable blotchy mottle appearance in the HLB-infected leaves [40, 70, 71]. The anatomical transverse section of HLB-infected leaf midrib exhibited phloem collapse with cell wall distortion and thickening in Valencia sweet orange and SB siblings [72]. In addition, hyperactive vascular cambium regenerates new phloem in the HLB-infected trees, consisting of assemblies of sieve elements, companion cells, and phloem parenchyma cells, but lacks phloemic fibers [72].

In addition to anatomical changes, several metabolic imbalances and genetic reprogramming are noticed in HLB-affected plants [57, 66]. Salicylic acid and

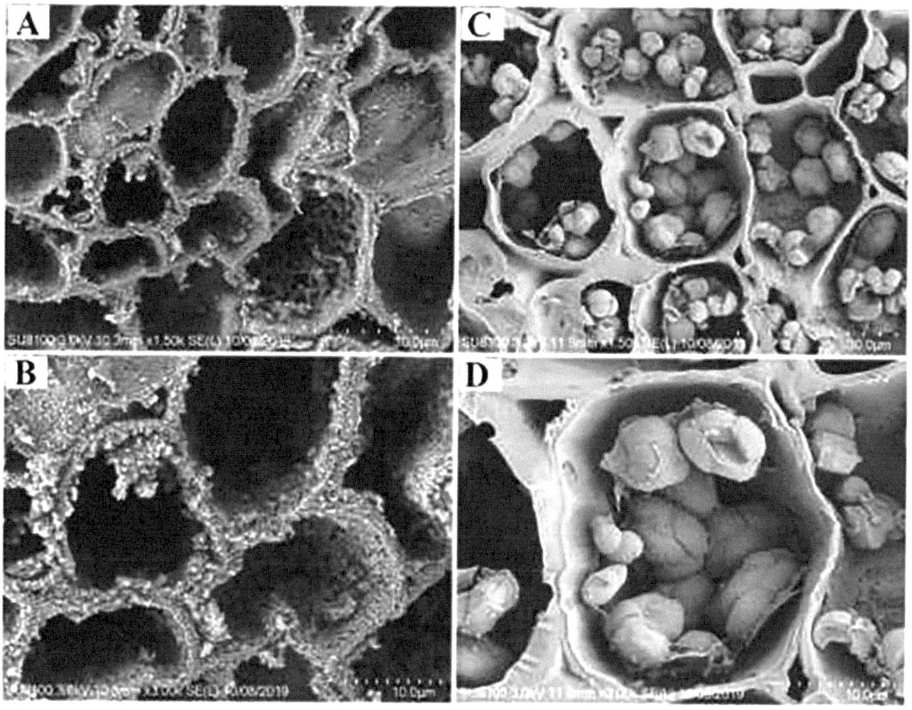

Figure 1.
SEM micrographs of transverse section of healthy and HLB-infected citrus petiole. *A and B. Healthy plant; C and D. HLB-infected plants.*

downstream signaling play a key role in provoking plant defense mechanisms against biotrophic pathogens [73, 74]. Wang and Trivedi postulated that a protein with potential salicylate hydroxylase activity might convert salicylic acid into catechol [75]. Salicylic acid pathway depression was observed in HLB-susceptible citrus plants [76]. Based on the *Candidatus* Liberibacter and plant interactions mechanism literature, we suggest the pathogenic mechanism of *Candidatus* Liberibacter species associated with citrus HLB in the following model (**Figure 2**).

4. Transmission of citrus HLB

The graft transmitted HLB was due to a viral disease [77]. Soon afterward, similar opinions were put forward in South Africa, strengthened by the results of grafting trials showing that greening was inconsistently transmitted to healthy plants. Lin [21] confirmed that HLB was transmitted through grafting in China, thus establishing the causative agent as a pathogen. McClean and Oberholzer [78] confirmed the graft transmissibility of African greening in 1965. The pathogen is not easily transmitted to progeny trees propagated by buds from infected trees, possibly due to sieve tube necrosis and uneven pathogen distribution, but more transmission occurs if stem pieces are used. No infection could be obtained when material from apparently healthy sectors of diseased trees was used. In 1964, natural spread by exposing seedlings to insects in a HLB-affected orchard developed yellowing symptoms similar to greening [79].

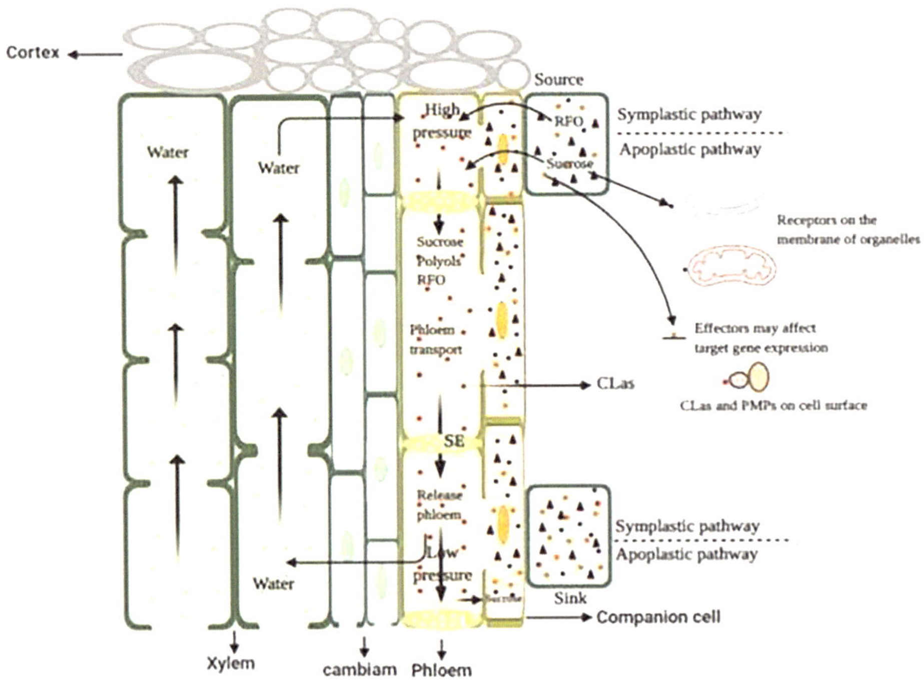

Figure 2.
Illustration of* Candidatus *Liberibacter virulence mechanisms in the plant. Candidatus *Liberibacter species associated with HLB (red circles) live in phloem elements. Phloem is mainly liable for the distribution of the carbohydrate from the source to the sink. Nutrients are transmitted to the phloem either through the apoplastic pathway or the symplastic pathway.* Candidatus *Liberibacter species may secrete effector protein (sec-dependent effectors) and virulence factors (orange and blue circle, red triangles) into phloem sieve elements and companion cells to interfere with host target (genes and protein) that can cause cell necrosis, cell death, and phloem malfunction. Effectors or virulence factors may interfere with phloem organelles, such as mitochondria, plastids, or endoplasmic reticulum, to trigger cellular responses. Some effectors (SDEs) may directly or indirectly affect the expression of target genes. In addition,* Candidatus *Liberibacter species may trigger plant immune responses through pathogen-associated molecular patterns leading to cell death and callose accumulation, resulting in inhibition of phloem transportation. The presence of* Candidatus *Liberibacter species and its metabolic activity may interfere with the function of the phloem by interrupting the osmatic gradients and integrity of phloem transportation. Abbreviations: PMPs—Pathogen-associated molecular patterns; RFO—Raffinose family oligosaccharide.*

Two insect vectors are responsible for the rapid transmission of citrus HLB from Las-infected citrus to healthy citrus species, Asian citrus psyllid *D. citri* in Asia and America, and the African citrus psyllid, *Triozaerytreae* in Africa. The acquisition feeding period is 30 min or longer, and the pathogen remains latent for 3–20 days. The inoculation feeding period is 1 hour or more [80].

Asian citrus psyllid is widespread around the world and found in hot and humid conditions and lower-lying areas in China, India, Myanmar, Taiwan, Philippine Islands, Malaysia, Indonesia, Sri Lanka, Pakistan, Thailand, Nepal, Ryukyu Islands (Japan), Afghanistan, Saudi Arabia, Reunion, and Mauritius [81]. Asian citrus psyllid firstly evolved in India [82], then spread in South America in the 1940s, invading Brazil, Argentina, and Venezuela, and then invaded the West Indies (Guadeloupe), Abaco Island, Grand Bahama Island, Cayman Islands, and the USA in the 1990s. In 2001, ACP was found in the Dominican Republic, Cuba, Puerto Rico, and Texas [83, 84]. Asian citrus psyllid has been reported more recently in many new Americas, including Mexico, Costa Rico, Belize, Honduras, and the states of Alabama, Arizona, California, Georgia, Louisiana, Mississippi, and South Carolina, USA [85].

African citrus psyllid (AfCP) thrives in cool and moist temperatures, at higher areas about 100 to 500 m above sea level. And it is sensitive to excessive heat and exists in Africa from the islands of the Indian Ocean through east and central Africa to South Africa, Saudi Arabia, Yemen, the northwestern region of the Iberian Peninsula, Cameroon, Kenya, Ethiopia, Zimbabwe, Tanzania, Malawi, Galicia, northern Portugal, Swaziland, Madagascar, Rwanda/Burundi, and Reunion [86]. Psyllid populations in Africa, Saudi Arabia, and Yemen might be able to adapt and settle under a wide range of environmental conditions, such as equatorial, arid, and warm temperate climates with varying temperatures and rainfall [86].

5. Current strategies to combat citrus HLB

5.1 Vector control

5.1.1 Biocontrol of vector

Biocontrol uses natural enemies by import, augmentation, and conservation to control the population density of disease pathogens or pests in agriculture [87]. Asian citrus psyllid (*Diaphorinacitri*) was controlled using a practical method of import (from Asia) and free of *Tamarixia radiata* in Florida citrus groves [88]. Natural enemies, *Diaphorencyrtus aligarhensis* and *Tamarixia radiata*, were imported from Taiwan and Vietnam to Florida citrus orchard and released as a biocontrol agent for *D. citri* [89]. *T. radiata* became more widely established parasitoid wasps than its counterpart *D. aligarhensis*. In urban and suburban regions, the release of *T. radiata* could considerably benefit commercial citrus growers by reducing latent psyllid populations and preventing the further spread of HLB disease [87]. *T. radiata* has been firmly established in commercial citrus-producing countries, including Réunion Island [90], the Philippines [91], Indonesia [92], Guadeloupe [93], and the USA [94], where it is spread throughout the state [88]. It appeared inadvertently in Brazil, Venezuela, Mexico [95], Puerto Rico [83], and Texas [31].

In addition to wasps, many insects native to Florida are recognized as *D. citri* predators, including several ladybeetle and spiders [81]. *Cyclonedas anguinea*, *Harmonia axyridis* Pallas (Coccinellid beetles), and *Olla v-nigrum* Mulsant were the leading killer of nymphal psyllids. In addition to *Ceraeochrysa*, *Hibanavelox* (becker), spider, and *Chrysoperla rufilabris* Burmeister (lacewings), *Tamarixia radiata* (parasitoid) also contributed to the additional mortality in Florida citrus groves. Coccinellid beetles are considered one of the most important natural enemies of *D. citri* populations in central Florida. Besides this, intraguild predation (IGP) causes more than 95% of immature *T. radiate* mortality [96]. Van den Berg et al. [97] noted that the spiders are the critical predators of *T. erytreae*, followed by chrysopids, coccinellids, syrphids, hemerobiids, Hemiptera, and predatory mites in citrus groves under the control of an integrated management program. Adults and larvae of *O. v-nigrum* (Mulsant) preying on premature Asian citrus psyllid were noticed in citrus groves throughout Florida [98].

A range of fungi species was reported to infect Asian citrus Psyllid, particularly under humid conditions [81], including entomopathogenic fungi, *Isaria fumosorosea*, *Lecanicillium lecanii*, *Beauveria bassiana*, *Capnodium citri*, *Cladosporium sp. nr. Oxysporum*, *Metarhizium anisopliae*, and *Hirsutella citriformis* that were used against HLB vectors as biopesticides [99–101]. *Isaria fumosorosea* is reported to have great potential to control different insect pests [102]. *H. citriformis* conidia with *synnemata* produced *in vitro* and *in vivo* were subjected to adult *D. citri* exhibits an increased mortality rate [100, 103]. In a 2-year field investigation in citrus groves in Florida, adult *D. citri* (ACP) was killed by *H. citriformis* following the rainy season [104]. In laboratory conditions, the fungal strains of *Isaria fumosorosea* (ESALQ-1296) and *Beauveria bassiana* (ESALQ-PL63) accounted for 77.8 and 78.4% of adult *D. citri* mortality, respectively, while in semifield conditions, adult *D. citri* mortality rate was as high as 83.5% with ESALQ-PL63 and 80.6% with ESALQ-1296. During 1 year, the monthly use of these two fungal strains in commercial citrus groves exhibited adult *D. citri* mortality ranging from 96.1% in December 2011 to 57.8% in October 2012. In addition, this study found that the mortality rate increased under high humidity conditions [99]. *Isariajavanica* and *Acrostalagmus aphidum* were also identified as biopesticides against *D. citri* in China [105]. The use of fungal species, such as *Metarhizium anisopliae*, *Cordyceps bassiana*, and *Isaria fumosorosea*, was shown to decrease larger populations of nymphs than adults of *D. citri* in the Persian lime groves [106].

5.1.2 RNA interference for vector control

RNA interference, a process in which a double-stranded RNA exerts a silencing effect on the complementary mRNA, has become a powerful tool in entomology. Advantages, such as ease of use, specific targeting, and lack of environmental persistence, make RNAi techniques highly attractive for crop protection against many insect pests [107]. The main challenges in using RNAi-based pest control methods are compelling target gene selection and reliable delivery of dsRNA. The overexpression of dsRNAs in transgenic plants has induced RNAi in targeted insects [108, 109]. The transgenic approach in citrus, however, is slow and difficult. Hajeri et al. [110] targeted *D. citri* endogenous *Awd* (abnormal wing development disc) gene for silencing by using a CTV-RNAi vector, resulting in impaired wings in *D. citri* that would potentially limit the ability to fly and successful transmission of CLas bacteria between citrus trees in the field. In addition, decreased *Awd* gene in nymphs resulted in malformed-wing phenotype in adults and increased adult mortality. Taning et al. [111] postulated that a small dose of dsRNA (dsAK, dsSOD) administered through in Planta system (iPS) bioassay was sufficient to trigger the RNAi mechanism, causing significant suppression of the targeted transcript and increased mortality in ACP.

5.1.3 Horticultural mineral oil for vector control

Petroleum-based horticultural mineral oils (HMO) are a vital constituent of integrated management programs for many pathogens and several phytophagous arthropods pathogens that affect the productivity of fruits, vegetables, and ornamentals in the commercial cultivation field as well as greenhouse conditions [112]. Since the 1980s, HMOs have been employed to control mites and scales in China [113]. HMO controls citrus leaf miner, citrus rust mite, citrus red mite, red scale, chaff scale, spiny whitefly, and Asian citrus Psyllid in citrus [114, 115]. By lowering the number of HMOs used in treatment to 0.25 ± 0.5% and maximizing the number of sprays during each season, a significant level of pest control was achieved without the threat of phytotoxicity. The combined treatment with oils and *Isaria fumosorosea* showed that the survival rate of adult psyllids was lower than that of oils used alone [116]. Kumar et al. [117] postulated that the combined treatment of entomopathogenic *Isaria fumosorosea* and HMOs dramatically reduced *D. citri* populations, where the maximum mean survival for *D. citri* was 12.5 ± 0.7 days. Similarly, Tansey et al. [118] revealed that mixes of insecticide and HMO application considerably decreased the populations of nymph and adult *D. citri* in Valencia orange groves in Florida. Conversely, Qureshi et al. [119] disclosed that HMO alone did not control *D. Citri* populations because the mean suppression of nymph and adults for more than 3 weeks was only 36 and 50%, respectively.

5.2 Las bacterial control

5.2.1 Antibiotics

Antibiotics are crucial for controlling bacterial diseases in fruit-bearing trees, vegetables, and ornamentals. Although antibiotics can be detected on plant surfaces using delicate analytical chemistry techniques for up to a month after application, their ability to inhibit bacterial growth is lost within a week [120]. In-plant disease control, nearly 40 antibiotics were screened; only streptomycin and tetracycline were used extensively in fruit trees [121]. The only commercially applied treatment for HLB was tetracycline, which is bacteriostatic rather than bactericidal, in Reunion Island's orchards [122, 123]. Tetracycline was the only approved antibiotic injection in trees injected directly into the trunks of HLB-affected citrus trees in China, Indonesia, India, Taiwan, and South Africa during the 1970s [36, 117, 124]. Although the symptoms of HLB were considerably decreased, this antibiotic trunk injection method was not in practice owing to its phytotoxicity and labor costs. The use of penicillin-carbendazim antibiotics in citrus trees showed significant control of HLB disease. The antibiotic disadvantage is a reduction in the fruit size owing to phytotoxicity and the residues of the antibiotics in citrus fruits [125]. The development of therapeutic compounds and bactericidal agents to control devastating HLB could provide an additional solution for an effective integrated disease management program. However, other than selective antibiotics, nonselective bactericide is recommended for general use in most crops, particularly citrus [126]. The combination treatment of streptomycin with penicillin efficiently eliminated or repressed the Las bacterium compared with the separate administration of either antibiotic [126]. The treatment of penicillin combined with streptomycin also significantly reduced the bacterial titer of Las in greenhouse citrus plants. Kasugamycin and Oxytetracycline combination therapy *via* trunk injection significantly reduced HLB bacterial titer in the field. However, the combination of kasugamycin and streptomycin was not effective against the bacterium of Las [127]. Penicillin with oxytetracycline combination therapy has been more effective in controlling citrus pathogens [128] but may require annual treatment [20]. Among the 31 tested antibiotics, some were effective at reducing and eliminating Las bacterial titers in inoculated rootstock and the treated scions of citrus plants, such as ampicillin, carbenicillin, penicillin, cefalexin, rifampicin, and sulfadimethoxine [20]. Oxytetracycline has therefore been suggested to be used more frequently in combination treatment [129, 130] with penicillin or kasugamycin against HLB to control the progression of bacterial resistance and maximize the antibiotic efficacy against HLB pathogenic bacteria [131]. The Environmental Protection Agency (EPA) of the USA allows citrus growers to spray streptomycin and oxytetracycline as routine treatments in the citrus field several times per year [132]. Oxytetracycline (1 g/L) was delivered to leaves of HLB-infected trees through the foliar application, and oxytetracycline was found in all leaves, although at reduced levels than in the directly applied leaves [132]. However, the phytotoxicity of tetracycline should be considered [20]. Antibiotics tested to combat HLB malady are tabulated in **Table 1**.

S.No	Antibiotics	Working concentration (mg/L)	Effectiveness	Phytotoxicity
1	Actidione	25	High	Highest
2	Validoxylamine A	100	Partly	Less
3	Zhongshengmycin	100	Partly	Less
4	Amikacin sulfate	100	None	NIL
5	Gentamicin sulfate	100	None	NIL
6	Hygromycin B	150	Partly	NIL
7	Kanamycin sulfate	100	Partly	None
8	Kasugamycin hydrochloride	100	None	NIL
9	Neomycin hydrate trisulfate	50	None	NIL
10	Spectinomycin dihydrochloride pentadrate	20	Partly	None
11	Streptomycin sulfate	100	None	NIL
12	Tobramycin	20	None	NIL
13	Ampicillin sodium	100	High	Less
14	Carbenicillin disodium	100	High	Less
15	Penicillin G potassium	100	High	Less
16	Cefalexin	100	High	Less
17	Vancomycin hydrochloride	40	None	NIL
18	Lincomycin hydrocloride	100	None	NIL
19	Cycloserine	50	Partly	NIL
20	Rifamycin sodium	50	Partly	Less
21	Rifampicin	50	High	Less
22	Rifaximin	50	Partly	Less
23	Colistinmethane sulfonate sodium	20	None	NIL
24	Polymixin B sulfate	300	None	NIL
25	Cinoxacin	300	None	NIL
26	Ciprofloxacin hydrochloride	300	Partly	NIL
27	Sulfadimethoxine sodium	100	Partly	Moderate
28	Sulfamethoxazole	100	Partly	Moderate
29	Sulfathiazole sodium	100	Partly	Moderate
30	Chloramphenicol	30	Partly	Less
31	Oxytetracycline hydrochloride	100	High	Highest

Table 1.
Antibiotics effectiveness against CLas bacterium and phytotoxicity.

5.2.2 Thermotherapy

Heat treatment or thermotherapy of planting material is a century-old disease control method that has proven effective against various pathogenic microorganisms. Thermotherapy, simple in principle, can eliminate the conserved pathogens depending on temperature/time regime and can cause mild injuries to the host during the treatment. Heat is mainly generated by water, vapor, or air [133]. The main advantage of thermotherapy treatment is that it is more environmentally friendly than harmful agrochemicals. Thermotherapy has proven to be an effective strategy against HLB that helps to enhance the vigor of citrus trees and promotes new root growth and development. The efficacy of thermotherapy against HLB pathogens depends on the temperature and citrus varieties [134]. Therapy could recuperate HLB-affected citrus plants by eliminating or suppressing Las bacterial titers at temperatures above 40°C [6, 134]. *Candidatus* Liberibacter asiaticus is a heat-tolerant phloem-limited bacteria that can withstand a temperature of about 35°C, while *Candidatus* Liberibacter americanus is heat-sensitive [135]. Thermotherapy could eliminate HLB pathogens from valuable horticultural trees associated with shoot tip grafting [136].

Lin opined on eliminating yellow shoot disease with water-saturated hot air treatment of graft wood 48–58°C with no loss of tissue viability [137]. In India, the thermotherapy of budwood at 47°C for 2 hours of diminished disease incidence, and more prolonged treatment eradicated the pathogen [138]. Heat treatment at temperatures around 38–40°C for 3 or 4 weeks killed HLB pathogens in young infected plants or citrus seedlings grafted with infected tissues [138, 139]. In South Africa, HLB-infected budwoods were treated with hot water baths at 51°C for 1 hour, 49°C for 2 hours, and 47°C for 4 hours, eliminating HLB pathogens with some loss of viability at higher temperatures [140]. In HLB-affected trees topped with polyethylene fiberglass sheets for 2 to 5 months, the number of diseased fruits decreased. However, this technique is not feasible for extensive use in citrus groves [27]. The HLB-affected citrus seedlings were continuously exposed to 40 to 42°C heat therapy for 7 to 10 days, significantly reducing titer or eliminating Las bacteria. This treatment can be helpful to combat HLB-affected plants in greenhouse and nursery settings [134]. Ehsani et al. [141] also postulated a decrease in HLB symptoms in groves of citrus trees after heat treatment. The combined thermo- and chemotherapy of sulfathiazole sodium or sulfadimethoxine sodium was more effective at 45°C than in thermotherapy alone, chemotherapy alone, or a combination of thermotherapy at 40°C and chemotherapy [142]. The temperature treatment at 45°C for 8 h per day for a week and a combination of ampicillin sodium, actidione, and validoxylamine A as a bark paint on grapefruits plant significantly reduced Las titer [143]. Two-year-old graft HLB-affected citrus reticulate treated with thermotherapy at 45°C and 48°C showed diminished HLB symptoms and Las titers 8 weeks after treatment in the greenhouse condition [144]. Commercial and residential citrus trees covered with portable plastic enclosures exposed to elevated temperatures through solarization showed vigorous growth in 3–6 weeks after treatment. Although commercial citrus trees showed Las after heat treatment, many trees generated extensive flushes and grew strongly for 2 to 3 years after therapy [145]. Inner bark heat treatment with 60°C–0.03 MPa-30s in 9-year-old citrus plants exhibited significantly reduced Las bacterial titer with vigorous plant growth from all treated HLB-affected trees [146]. Abdulridha et al. [147] reported that HLB-affected trees with canopy cover were treated with combined hot water and steam therapy at 55°C for 90 seconds. The temperature distribution inside the canopy cover was not uniform; the canopy temperatures were more significant than the trunk temperatures. The mobile thermotherapy treatment needs to be improved to increase the temperatures around the tree trunk to nearly the same temperature as a canopy. Vincent et al. [132] postulated that heat treatment from 43 to 54°C for no longer than 45 s showed adverse effects on citrus tree growth.

HLB is a systemic disease. Efficient elimination of Las bacteria from the entire citrus tree, including roots, is vital to managing the disease. The current thermotherapy challenge is that although adequately elevated temperatures can reach the above-ground areas of the plant, killing temperatures are unlikely to be attained at the roots where the temperature is mitigated by the soil [148]. Therefore, heat treatment is unlikely to reduce the populations of HLB pathogens in the roots, which then acts as a site for canopy reinfection during flushes. The efficacy of heat treatment in eliminating Las bacterial populations in underground roots must be enhanced to become a feasible part of integrated citrus HLB management [15]. To overcome this barrier, Hoffman et al. [134] suggest that heat treatment, coupled with chemotherapy in HLB-affected plants, can lead to a potential future strategy for controlling citrus HLB.

5.2.3 Plant defense activators to combat HLB

Trunk injection is an alternative target-precise technique for efficiently delivering plant protective chemicals in tree fruit crops. It harnesses the rapid transportation ability of the xylem that enables therapeutic compounds' translocation and subsequent distribution into the canopy where plant protection is needed [149]. There has been limited research on the trunk injection of antibiotics and plant defense activators for better disease control. Several recent field studies have demonstrated the utility of trunk injection of bactericides and plant defense activators in disease management [150].

Treatments with β-aminobutyric acid (BABA), 2,1,3-benzothiadiazole (BTH), 2,6-dichloroisonicotinic acid (INA), ascorbic acid (AA), and the nonmetabolizable glucose analog 2-deoxy-D-glucose (2-DDG) plant defense inducers individually or in combination found effective in suppressing Las bacterial population in plants and sustaining fruit production to a certain extent. Treatment with BABA and BTH was the most effective in reducing the Las population in plant tissues compared with other plant defense inducers [151]. Hu and Wang proved that trunk injection of oxytetracycline in HLB-affected trees exhibited long-lasting suppression of Las populations. It also prevented the tree decline by promoting new growth without the disease [152]. Trunk injections of salicylic acid, potassium phosphate, acibenzolar-S-methyl, and oxalic acid in the HLB-affected tree significantly suppressed the Las titer and HLB disease progress [150].

Brassinosteroids (BRs) are a class of steroid hormones that regulate gene expression, growth, and developmental processes in response to biotic and abiotic stress [153]. The plant defense mechanism of brassinosteroids was mediated by leucine-rich repeat receptor kinase (LRR-RK) BAK1, which serves as a coreceptor for both microbe-associated molecular patterns (MAMPs) and steroid hormone [154], which binds to BRs and FLS2 eliciting microbe-induced immunity. BR treatment showed increasing disease resistance against many pathogens [6]. Canales et al. [155] postulated that applying epibrassinolide as a foliar spray in HLB-infected plants improved immunity against *Candidatus* Liberibacter asiaticus in greenhouse and field citrus plants. *Candidatus* Liberibacter asiaticus titer was markedly reduced in epibrassinolide-treated plants due to the enhanced defense gene expression in the citrus leaves. However, the molecular mechanism of BRs in plant responses under normal and environmentally challenging conditions has remained unclear [155].

5.3 Nanoemulsions to deliver chemicals against Las bacteria

HLB is caused by Las proteobacteria that reside in the phloem of infected citrus trees. It is, therefore, challenging to deliver effective compounds into the phloem through a foliar spray. The presence of wax, cutin, and pectin in plant cuticles prevents the effective bactericidal compounds from entering the phloem through a foliar spraying method. The use of chemical adjuvant enhanced the foliar uptake of agrochemicals [156, 157]. However, foliar spray treatment, including the combination of antibiotic PS and adjuvants in dimethyl sulfoxide and Silwet L-77, did not significantly impact the HLB-affected citrus trees [128]. Therefore, there is a need for candidate adjuvants, which can potentially increase the permeability of citrus cuticles to deliver antimicrobial compounds into citrus phloem.

Nanoemulsions or submicron emulsions are colloidal dispersion systems with average droplets size ranging from 50 to 1000 nm that has extensively studied for delivering chemical compounds. Nanoemulsions were pondered as thermodynamically and kinetically stable isotropic dispersions, composed of two immiscible liquids such as water and oil, stabilized by an interfacial film composed of an appropriate surfactant and co-surfactant to form a single-phase [158]. However, the approach efficacy relies on nanoemulsions droplet characteristics, such as low surface tension, tiny size, ample surface area, and low interface tension [159]. Our research group postulated that water in oil nanoemulsions containing ampicillin coupled with adjuvant Brij 35 was used as a foliar spray to enhance the permeability through the citrus cuticle into the phloem and more efficiently eliminated Las bacteria in HLB-affected citrus *in planta* [160]. Ampicillin showed the lowest phytotoxicity to citrus trees infected with Las bacteria [20]. However, the US Environmental Protection Agency (EPA) has not approved the commercial use of ampicillin in crops due to the development of resistant bacterial strains [160]. In another study, oil in water nanoemulsions was formulated using a spontaneous emulsification method, where five different antimicrobial compounds alone combined with Cremophor EL (viscous oil), acetone, and Span 80/Tween 80, which formed tiny droplets, were effectively applied to the bark for efficiently control HLB [161].

Silver nanoparticles (AgNPs) are one of the most investigated and used in agricultural science to enhance the yield and sustainable development of the crop. This has long been reported to have significant antibacterial, antifungal, antiviral, and pesticide effects. AgNPs are used as foliar sprays to prevent the development of rot, mold, fungi, and other plant pathogens [162]. Stephano-Hornedo et al. [18] evaluated the commercially available AgNPs to directly eradicate *Candidatus* Liberibacter asiaticus (CLas), responsible for HLB in the citrus field. The 93 HLB-infected citrus trees administered foliar and trunk injections of silver nanoparticles showed a remarkable reduction of 80–90% in bacterial titer by both methods than control. Compared with other effective treatments involving b-lactam antibiotics, the effectiveness of AgNPs is 3- to 60-fold higher when administered by foliar spray and 75- to 750-fold higher when injected *via* tree trunk. Thus, the silver nanoparticles could be a sustainable method for mitigating citrus HLB. However, AgNPs toxicity to a citrus tree and the environment needs to be warranted before its commercial use.

5.4 Transgenic approach to combat HLB

Globally, insect pests are responsible for significant crop losses through direct harm and transmission of plant diseases [163]. The best long-term alternative strategy for managing citrus HLB is to develop disease-resistant cultivars in commercial citrus production. Due to the lack of resistant cultivars, developing HLB-resistant plants by conventional citrus breeding is difficult. Resistance occurs in citrus relatives, such as kumquat, where its genetic background influences the quality and yield of the fruit [164]. In addition, conventional citrus breeding is labor- and time-consuming, and very costly as citrus species are polygenic, extremely heterozygous plants with a long juvenile phase. The genetic transformation approach is an essential strategy that would aid in incorporating disease-resistant genes into citrus cultivars to combat the HLB disease. The progression of citrus breeding through genetic transformation is still early, indicating a lack of molecular pathogenesis understanding of innate disease resistance in citrus [165].

Systemic acquired resistance (SAR), a natural plant defense response mechanism, has been well characterized in *Arabidopsis thaliana*. SAR entails signal molecule salicylic acid (SA) to activate defense mechanisms. In response to SA, the non-expression of pathogenesis-related gene 1 (NPR1) is translocated to the nucleus, where it triggers the expression of pathogenic related (PR) genes by interacting with TGA transcription factors, thereby provoking SAR [166, 167]. *Arabidopsis* mutants contain deficiencies in the *NPR1* gene showing decreased PR gene expression induced by SA and SAR, leading to increased susceptibility to pathogens [167, 168]. Conversely, overexpression of the *NPR1* gene in *Arabidopsis* increased the disease resistance to bacteria and oomycete pathogens. Interestingly, the over-expression of *AtNPR1*gene in most plant species does not provoke noticeable adverse effects on plant growth and development [169]. Thus, *NPR1* is a target gene for the genetic transformation of nonspecific resistance in crop plants.

Dutt et al. [170] postulated that the overexpression of the *AtNPR1* gene in Hamlin and Valencia orange cultivars resulted in trees with normal phenotypes, and exhibited increased resistance to HLB. Transgenic trees showed reduced disease severity, and a few lines remained disease-free even after 36 months of planting in a high-disease pressure field. The phloem-expressed *NPR1* gene was equally effective in increasing disease resistance by triggering several indigenous gene expressions involving plant defense mechanisms of signaling pathways. In addition to triggering resistance to HLB, the observed SAR response could protect citrus trees from other major fungal and bacterial diseases, such as black spots and citrus canker [170].

6. Conclusions

HLB is one of the century-old diseases in the history of citrus pathology. The global spread of HLB disease causes economic loss in most citrus-producing countries. The causal agent of HLB, *Candidatus* Liberibacter, impedes understanding its pathogenic mechanism, fastidious nature, and unculturable in artificial conditions. Future research will focus on the isolation and pure culture of these proteobacteria. The rootstock and scion of some tolerant varieties have been noted and used by citrus growers in citriculture. These tolerant varieties have the potential to suppress the progression of HLB in Las-infected trees. The HLB management program recommends the intense psyllid control and removal of Las-infected trees in citrus groves. Citrus farmers focus on maintaining productivity in the HLB-affected trees by using plant defense activators, micronutrients, and fertilizers and paying more attention to water irrigation systems. Besides, thermotherapy is still an efficient methodology for eliminating and suppressing the causal agents in citrus scions and rootstock.

Antibiotics alone or combined with other bactericides have also shown to be effective against citrus HLB. However, antibiotics need to evaluate their phytotoxicity before their commercial use. The combination of thermotherapy and antibiotics, plant defense activators, and thermotherapy provides controlled HLB management efficiency. The formation of nanoemulsion in water in oil (W/O) and/or oil in water (O/W) could offer a practical methodology for the targeted delivery of antimicrobial compounds to the phloem of citrus by foliar spraying method to control citrus HLB. In addition, transgenic orange cultivars over-expressing the *AtNPR1* gene exhibited enhanced resistance to HLB. Transcriptome analysis between susceptible, tolerant, and resistant citrus varieties provides new insights into HLB tolerance by revealing defense-related genes, biological pathway signaling, hormones, transporters, carbohydrate metabolism, phloem-related genes, and secondary metabolism. In addition, some potential targets have also been identified, such as *DMR6-like* and *NPR1-like* genes for future HLB-tolerant citrus breeding [171]. Epibrassinolide as a foliar spray in HLB-affected plants improved the immunity against *Candidatus* Liberibacter asiaticus in the greenhouse and citrus field. However, further studies on the impact of eBL and nanoemulsion loaded with antibiotics in HLB-affected citrus plants are warranted to understand the complexity of citrus pathophysiology and fruit productivity. Researchers have investigated many control strategies to combat *Candidatus* Liberibacter species, but no effective management strategies have been developed. More studies are needed to investigate a sustainable and environmentally friendly strategy to control citrus HLB in the form of an antimicrobial agent in citrus groves. Meanwhile, biotechnological approaches such as transgenic, gene editing, and host-induced gene silencing may provide an unprecedented opportunity for long-term HLB management tools.

Based on the extensive prevention strategy experiments in the citriculture field by Chinese farmers, it has been shown that the control of HLB disease can be carried out in the three-pronged approach.

1. **HLB-free seedlings.** Selection of HLB-free citrus saplings, rootstocks, and scions. Furthermore, infected root stocks/scions might be cured through thermotherapy.

2. **Removal of infected plants.** Identification of HLB-affected plants by utilizing a suitable pathogen detection system and removing infected trees or infected sectors.

3. **Suppress the psyllid.** Control of psyllids to reduce the spread of HLB pathogens in the field and biological control of vectors might be desirable methods to control the vector populations rapidly and cost-effectively.

Nanotechnology-driven farming is still early, but it is an exciting and challenging field of research to be developed in the future, especially if the proper emphasis is placed on understanding the fundamental interactions between nanoscale materials and crop plants [172]. Future nanotechnology will enable the development of biosensors for early diagnosis of disease, new methods for suppression of disease pathogens in field and greenhouse conditions, and new molecular tools for understanding pathogenic mechanisms in pathogens and plants [173]. Nanotechnological investigations in phytopathology have increased dramatically over the last decade. Nanomaterials can be engineered as biosensors to diagnose plant diseases and as a means of delivery of genetic material, probes, and agrochemicals. Nanotechnology has been incorporated into disease management strategies, diagnostic tools, and molecular tools. Nanotechnologies could provide an alternative treatment to citrus farmers to be integrated into their existing HLB management programs in the citrus groves.

Acknowledgements

This work was funded by the Science and Technology Major Project of Guangxi (Gui Ke AA18118046).

Conflict of interest

The authors declare no conflict of interest.

Author details

Palaniyandi Karuppaiya, Junyuan Huang and Muqing Zhang*
State Key Laboratory for Conservation and Utilization of Subtropical Agro-bioresources, Guangxi University, Nanning, China

*Address all correspondence to: mqzhang@ufl.edu; zmuqing@163.com

References

[1] Liu Y, Heying E, Tanumihardjo SA. History, global distribution, and nutritional importance of citrus fruits. Comprehensive Reviews in Food Science and Food Safety. 2012;**11**(6):530-545

[2] Vashisth T, Chun C, Hampton OH. Florida Citrus nursery trends and strategies to enhance production of field-transplant ready Citrus plants. Horticulturae. 2020;**6**(1):8

[3] Mahato N, Sharma K, Sinha M, Baral ER, Koteswararao R, Dhyani A, et al. Bio-sorbents, industrially important chemicals and novel materials from citrus processing waste as a sustainable and renewable bioresource: A review. Journal of Advanced Research. 2020;**23**:61-82

[4] Castle WS, Krezdorn AH. Rootstocks for Florida Citrus: Rootstock Selection, the First Step to Success. Institute of Food and Agricultural Sciences. FL, USA: University of Florida; 1993

[5] Stover E, Inch S, Richardson ML, Hall DG. Conventional citrus of some scion/rootstock combinations show field tolerance under high Huanglongbing disease pressure. HortScience. 2016;**51**(2):127-132

[6] Munir S, He P, Wu Y, He P, Khan S, Huang M, et al. Huanglongbing control: Perhaps the end of the beginning. Microbial Ecology. 2018;**76**(1):192-204

[7] Da Graça J, Korsten L. Citrus huanglongbing: Review, present status, and future strategies. In: Diseases of Fruits and Vegetables. Vol. I. Dordrecht: Springer; 2004. pp. 229-245

[8] Chung K, Brlansky R. Citrus diseases exotic to Florida: Huanglongbing (citrus greening). In: Plant Pathology Department Fact Sheet PP-210. Florida Cooperative Extension Service, Institute of Food and Agricultural Sciences. FL, USA: University of Florida; 2005. Available from: http://edis.ifas.ufl.edu/PP133

[9] Jagoueix S, Bove JM, Garnier M. The phloem-limited bacterium of greening disease of citrus is a member of the alpha subdivision of the Proteobacteria. International Journal of Systematic Bacteriology. 1994;**44**(3):379-386

[10] Zhao XY. Citrus yellow shoot disease (Huanglongbing) in China—A review. Proceedings of the International Society of Citriculture. 1981;**1**:466-469

[11] Bové JM. Huanglongbing: A destructive, newly-emerging, century-old disease of citrus. Journal of Plant Pathology. 2006;**88**(1):7-37

[12] Albrecht U, Bowman KD. Tolerance of the trifoliate citrus hybrid US-897 (Citrus reticulata Blanco× Poncirus trifoliata L. Raf.) to Huanglongbing. HortScience. 2011;**46**(1):16-22

[13] Omura M, Shimada T. Citrus breeding, genetics, and genomics in Japan. Breeding Science. 2016;**66**(1):3-17

[14] Yang C, Zhong Y, Powell CA, Doud MS, Duan Y, Huang Y, et al. Antimicrobial compounds effective against *Candidatus* Liberibacter asiaticus discovered via graft-based assay in Citrus. Scientific Reports. 2018;**8**(1):17288

[15] Wang N, Pierson EA, Setubal JC, Xu J, Levy JG, Zhang Y, et al. The *Candidatus* Liberibacter-host interface: Insights into pathogenesis

mechanisms and disease control. Annual Review of Phytopathology. 2017;**55**:451-482

[16] Ghosh D, Motghare M, Gowda S. Citrus greening: Overview of the most severe disease of Citrus. Advanced Agricultural Research & Technology Journal. 2018;**2**(1):83-100

[17] Blaustein RA, Lorca GL, Teplitski M. Challenges for managing *Candidatus* Liberibacter spp. (Huanglongbing disease pathogen): Current control measures and future directions. Phytopathology. 2018;**108**(4):424-435

[18] Stephano-Hornedo JL, Torres-Gutiérrez O, Toledano-Magaña Y, Gradilla-Martínez I, Pestryakov A, Sánchez-González A, et al. Argovit™ silver nanoparticles to fight Huanglongbing disease in Mexican limes (*Citrus aurantifolia* Swingle). RSC Advances. 2020;**10**:6146-6155

[19] Zhao XY. Citrus yellow shoot disease (Huanglongbing) in China—A review. Proceedings of the International Society of Citriculture. 1981;**1**:466-469

[20] Zhang M, Guo Y, Powell CA, Doud MS, Yang C, Duan Y. Effective antibiotics against '*Candidatus* Liberibacter asiaticus' in HLB-affected citrus plants identified via the graft-based evaluation. PLoS One. 2014;**9**(11):e111032

[21] Lin KH. Observations on the yellow shoot of citrus. Etiological studies of the yellow shoot of citrus. Acta Phytopathologica Sinica. 1956;**2**(1):1-42

[22] Capoor S. Decline of citrus trees in India. Bulletin of the National Institute of Sciences of India. 1963;**24**:48-64

[23] Raychaudhuri S. et al. Recent studies on citrus greening in India. International Organization of Citrus Virologists Conference Proceedings (1957-2010). 1974;**6**(6)

[24] Obergolzer P, Von Standen D, Basson W. Greening disease of sweet orange in South Africa. International Organization of Citrus Virologists Conference Proceedings (1957-2010). 1965;**3**(3)

[25] Aubert B et al. Greening, a Serious Threat for the Citrus Productions of the Indonesian Archipelago. Future Prospects of Integrated Control. Fruits (France): IRFA; 1985

[26] Moreno P, da Graça J, Yokomi R. Preface. In: Proceedings of the 13th Conference of the International Organization of Citrus Virologists. Riverside: IOCV; 1996

[27] Graca JD. Citrus greening disease. Annual Review of Phytopathology. 1991;**29**(1):109-136

[28] Lee HA. The relation of stocks to the mottled leaf of citrus trees. Philippine Journal Science. 1921;**18**:85-95

[29] Van der Merwe A. Chromium and manganese toxicity. Is it important in Transvaal citrus greening? Farming in South Africa. 1937;**12**:439-440

[30] McCollum G, Baldwin E. Huanglongbing devastating disease of Citrus. Horticultural Reviews. 2016;**44**:315-361

[31] Dala-Paula BM, Plotto A, Bai J, Manthey JA, Baldwin EA, Ferrarezi RS, et al. Effect of Huanglongbing or greening disease on Orange juice quality, a review. Frontiers in Plant Science. 2019;**9**:1976

[32] Schneider H. Anatomy of greening-diseased sweet orange shoots. Phytopathology. 1968;**58**(1):1555-1160

[33] Koen T, Langenegger W. Effect of the greening virus on the macro-element content of citrus leaves. Farming in South Africa. 1970;**45**(12)

[34] McClean A, Schwarz R. Greening or blotchy-mottle disease of citrus. Phytophylactica. 1970;**2**(3):177-194

[35] Gottwald TR. Citrus canker and citrus huanglongbing, two exotic bacterial diseases threatening the citrus industries of the Western Hemisphere. Outlooks on Pest Management. 2007;**18**(6):274-279

[36] Van Vuuren S. The determination of optimal concentration and pH of tetracycline hydrochloride for trunk injection of greening-infected citrus trees. Phytophylactica. 1977;**9**(4):77-82

[37] Korsten L et al. Detection of citrus greening-infected citrus in South Africa using a DNA probe and monoclonal antibodies. International Organization of Citrus Virologists Conference Proceedings (1957-2010). 1993;**12**(12)

[38] Graham J, Menge J. Phytophthorainduced diseases. In: Compendium of Citrus Diseases. MN, USA: APS Press; 2000:12-15

[39] Kogenaru S, Yan Q, Riera N, et al. Repertoire of novel sequence signatures for the detection of *Candidatus* Liberibacter asiaticus by quantitative real-time PCR. BMC Microbiology. 2014;**14**:39

[40] Etxeberria E, Gonzalez P, Achor D, Albrigo G. Anatomical distribution of abnormally high levels of starch in HLB-affected Valencia orange trees. Physiological and Molecular Plant Pathology. 2009;**74**(1):76-83

[41] Shokrollah H, Abdullah TL, Sijam K, Abdullah SNA. Potential use of selected citrus rootstocks and interstocks against HLB disease in Malaysia. Crop Protection. 2011;**30**(5):521-525

[42] Arredondo Valdés R, Delgado Ortiz JC, Beltrán Beache M, Anguiano Cabello J, Cerna Chávez E, Rodríguez Pagaza Y, et al. A review of techniques for detecting Huanglongbing (greening) in citrus. Canadian Journal of Microbiology. 2016;**62**(10):803-811

[43] Pereira FMV, Milori DM. Investigation of the stages of citrus greening disease using micro synchrotron radiation X-ray fluorescence in association with chemometric tools. Journal of Analytical Atomic Spectrometry. 2010;**25**(3):351-355

[44] Tran TT, Clark K, Ma W, Mulchandani A. Detection of a secreted protein biomarker for citrus Huanglongbing using a single-walled carbon nanotubes-based chemiresistive biosensor. Biosensors & Bioelectronics. 2020;**147**:111766

[45] Iftikhar Y, Rauf S, Shahzad U, Zahid MA. Huanglongbing: Pathogen detection system for integrated disease management–a review. Journal of the Saudi Society of Agricultural Sciences. 2016;**15**(1):1-11

[46] Davis MJ, Mondal SN, Chen H, Rogers ME, Brlansky RH. Co-cultivation of '*Candidatus* Liberibacter asiaticus' with Actinobacteria from Citrus with Huanglongbing. Plant Disease. 2008;**92**(11):1547-1550

[47] Folimonova SY, Achor DS. Early events of citrus greening (Huanglongbing) disease development at the ultrastructural level. Phytopathology. 2010;**100**(9):949-958

[48] Sechler A, Schuenzel EL, Cooke P, Donnua S, Thaveechai N, Postnikova E,

et al. Cultivation of '*Candidatus* Liberibacter asiaticus', '*Ca*. L. africanus', and '*Ca*. L. americanus' associated with huanglongbing. Phytopathology. 2009;**99**(5):480-486

[49] Hilf ME. Colonization of citrus seed coats by '*Candidatus* Liberibacter asiaticus': Implications for seed transmission of the bacterium. Phytopathology. 2011;**101**(10):1242-1250

[50] Green ER, Mecsas J. Bacterial secretion systems: An overview. Microbiology Spectrum. 2016;**4**(1):10.1128

[51] Shi Q, Pitino M, Zhang S, Krystel J, Cano LM, Shatters RG Jr, et al. Temporal and spatial detection of *Candidatus* Liberibacter asiaticus putative effector transcripts during interaction with Huanglongbing-susceptible, –tolerant, and -resistant citrus hosts. BMC Plant Biology. 2019;**19**(1):122

[52] Duan Y, Zhou LJ, Hall DG, et al. Complete genome sequence of citrus huanglongbing bacterium,'*Candidatus* Liberibacter asiaticus' obtained through metagenomics. Molecular Plant-Microbe Interactions. 2009;**22**(8):1011-1020

[53] Sengoda VG, Munyaneza JE, Crosslin JM, et al. Phenotypic and etiological differences between psyllid yellows and Zebra Chip diseases of potato. American Journal of Potato Research. 2010;**87**:41-49

[54] Katoh H, Miyata S, Inoue H, Iwanami T. Unique features of a Japanese '*Candidatus* Liberibacter asiaticus' strain revealed by whole genome sequencing. PLoS One. 2014;**9**(9):e106109

[55] Hao G, Boyle M, Zhou L, Duan Y. The intracellular citrus huanglongbing bacterium, '*Candidatus* Liberibacter asiaticus' encodes two novel autotransporters. PLoS One. 2013;**8**(7):e68921

[56] Li J, Pang Z, Trivedi P, Zhou X, Ying X, Jia H, et al. '*Candidatus* Liberibacter asiaticus' encodes a functional salicylic acid (SA) hydroxylase that degrades SA to suppress plant defenses. Molecular Plant-Microbe Interactions. 2017;**30**(8):620-630

[57] Dalio RJD, Magalhães DM, Rodrigues CM, Arena GD, Oliveira TS, Souza-Neto RR, et al. PAMPs, PRRs, effectors and R-genes associated with citrus-pathogen interactions. Annals of Botany. 2017;**119**(5):749-774

[58] Zipfel C, Robatzek S. Pathogen-associated molecular pattern-triggered immunity: Veni, vidi...? Plant Physiology. 2010;**154**(2):551-554

[59] Zou H, Gowda S, Zhou L, Hajeri S, Chen G, Duan Y. The destructive citrus pathogen, '*Candidatus* Liberibacter asiaticus' encodes a functional flagellin characteristic of a pathogen-associated molecular pattern. PLoS One. 2012;7(9):e46447

[60] Wulff NA, Zhang SJ, Setubal JC, et al. The complete genome sequence of '*Candidatus* Liberibacter americanus', associated with citrus huanglongbing. Molecular Plant-Microbe Interactions. 2014;**27**(2):163-176

[61] Yan Q, Sreedharan A, Wei S, Wang J, Pelz-Stelinski K, Folimonova S, et al. Global gene expression changes in *Candidatus* Liberibacterasiaticus during the transmission in distinct hosts between plant and insect. Molecular Plant Pathology. 2013;**4**:391-404

[62] Zhang SJ, Flores-Cruz Z, Zhou LJ, et al. '*Ca*. Liberibacter asiaticus' carries an excision plasmid prophage and a chromosomally integrated prophage

that becomes lytic in plant infections. Molecular Plant-Microbe Interactions. 2011;**24**(4):458-468

[63] Zheng Z, Bao M, Wu F, Van Horn C, Chen J, Deng X. A type 3 prophage of '*Candidatus* Liberibacter asiaticus' carrying a restriction-modification system. Phytopathology. 2018;**108**(4):454-461

[64] Beaulaurier J, Schadt EE, Fang G. Deciphering bacterial epigenomes using modern sequencing technologies. Nature Reviews. Genetics. 2019;**20**(3):157-172

[65] Zheng Z, Bao M, Wu F, Chen J, Deng X. Predominance of single prophage carrying a CRISPR/cas system in "*Candidatus* Liberibacter asiaticus" strains in southern China. PLoS One. 2016;**11**(1):e0146422

[66] Aritua V, Achor D, Gmitter FG, Albrigo G, Wang N. Transcriptional and microscopic analyses of citrus stem and root responses to *Candidatus* Liberibacter asiaticus infection. PLoS One. 2013;**8**(9):e73742

[67] Koh EJ, Zhou L, Williams DS, Park J, Ding N, Duan YP, et al. Callose deposition in the phloem plasmodesmata and inhibition of phloem transport in citrus leaves infected with "*Candidatus* Liberibacter asiaticus". Protoplasma. 2012;**249**(3):687-697

[68] Albrigo L et al. Phloem disruption from HLB infection in canopy and root framework. Journal of Citrus Pathology. 2014;**1**(1)

[69] Etxeberria E, Narciso C. Anatomy of the citrus leaf petiole: Healthy vs. Huanglongbing. Acta Horticulturae. 2015;**1065**:891-897

[70] Gonzalez P, Reyes-De-Corcuera J, Etxeberria E. Characterization of leaf starch from HLB-affected and unaffected-girdled citrus trees. Physiological and Molecular Plant Pathology. 2012;**79**:71-78

[71] Achor D. Citrus affected with huanglongbing disease. Plant Pathology Journal. 2010;**9**:56-64

[72] Deng H, Achor D, Exteberria E, Yu Q, Du D, Stanton D, et al. Phloem regeneration is a mechanism for Huanglongbing-tolerance of "Bearss" lemon and "LB8-9" sugar belle® mandarin. Frontiers in Plant Science. 2019;**10**:277

[73] An C, Mou Z. Salicylic acid and its function in plant immunity. Journal of Integrative Plant Biology. 2011;**53**(6):412-428

[74] Fu ZQ, Dong X. Systemic acquired resistance: Turning local infection into global defense. Annual Review of Plant Biology. 2013;**64**(1):839-863

[75] Wang N, Trivedi P. Citrus huanglongbing: A newly relevant disease presents unprecedented challenges. Phytopathology. 2013;**103**(7):652-665

[76] Xu M, Li Y, Zheng Z, Dai Z, Tao Y, Deng X. Transcriptional analyses of Mandarins seriously infected by '*Candidatus* Liberibacter asiaticus'. PLoS One. 2015;**10**(7):e0133652

[77] Chen Q. A report of a study on yellow shoot of citrus in Chaoshan. New Agricultural Quarterly Bulletin. 1943;**3**(142):75

[78] McClean A, Oberholzer P. Greening disease of the sweet orange: Evidence that it is caused by a transmissible virus. South African Journal of Agricultural Science. 1965;**8**(1)

[79] Schwarz R. An insect-transmissible virus, trapped on sweet orange seedlings

in orchards where greening disease is common-research note. South African Journal of Agricultural Science. 1964;7(4):885-890

[80] Huang C et al. Incidence and spread of citrus likubin in relation to the population fluctuation of *Diaphorina citri*. Plant Protection Bulletin (Taipei). 1990;**32**(3):167-176

[81] Grafton-Cardwell EE, Stelinski LL, Stansly PA. Biology and management of Asian citrus psyllid, vector of the huanglongbing pathogens. Annual Review of Entomology. 2013;**58**:413-432

[82] Beattie G et al. On the origins of Citrus, Huanglongbing, *Diaphorina citri* and Trioza erytreae. In: Proceedings of the International Research Conference on Huanglongbing. Orlando: Florida Citrus Mutual; 2008

[83] Pluke RW, Qureshi JA, Stansly PA. Citrus flushing patterns, *Diaphorina citri* (Hemiptera: Psyllidae) populations and parasitism by *Tamarixia radiata* (Hymenoptera: Eulophidae) in Puerto Rico. The Florida Entomologist. 2008;**91**(1):36-42

[84] French J, Kahlke C, Da Graça J. First record of the Asian citrus psylla, *Diaphorina citri* Kuwayama (Homoptera: Psyllidae) in Texas. Subtropical Plant Science. 2001;**53**:14-15

[85] Hall DG, Richardson ML, El-Desouky Ammar ED, Halbert SE. Asian citrus psyllid, *Diaphorina citri*, vector of citrus huanglongbing disease. Entomologia Experimentalis et Applicata. 2013;**146**(2):207-223

[86] Cocuzza GEM, Alberto U, Hernández-Suárez E, et al. A review on *Trioza erytreae* (African citrus psyllid), now in mainland Europe, and its potential risk as vector of huanglongbing (HLB) in citrus. Journal of Pest Science. 2017;**90**:1-17

[87] Alvarez S, Rohrig E, Solís D, et al. Citrus greening disease (Huanglongbing) in Florida: Economic impact, management and the potential for biological control. Agricultural Research. 2016;**5**:109-118

[88] Qureshi JA et al. Incidence of invasive *Diaphorina citri* (Hemiptera: Psyllidae) and its introduced parasitoid *Tamarixia radiata* (Hymenoptera: Eulophidae) in Florida citrus. Journal of Economic Entomology. 2009;**102**(1):247-256

[89] Rohrig EA et al. Field release in Florida of *Diaphorencyrtus aligarhensis* (Hymenoptera: Encyrtidae), an endoparasitoid of *Diaphorina citri* (Homoptera. Psyllidae), from mainland China. The Florida Entomologist. 2012;**95**(2):479-481

[90] Aubert B, Quilici S. Biological control of the African and Asian citrus psyllids (Homoptera: Psylloidea), through eulophid and encyrtid parasites (Hymenoptera: Chalcidoidea) in Reunion Island. International Organization of Citrus Virologists Conference Proceedings (1957-2010). 1984;**9**(9)

[91] Gavarra M et al. Progress report: *D. citri* trapping, identification of parasite and possible field establishment of the imported parasite, *Tamarixia radiata* in the Philippines. In: Proceedings of the Fourth International Asia Pacific Conference on Citrus Rehabilitation, Chiang Mai, Thailand. 1990. pp. 4-10

[92] Nurhadi F. Records of Important Parasites Attacking *Diaphorina citri* in East Java, Indonesia. Fuzhou, China: Workshop Citrus Greening Disease, Huanglungbin, FAO-UNDP Project; 1987

[93] Étienne J, Quilici S, Marival D, Franck A. Biological control of *Diaphorina citri* (Hemiptera: Psyllidae) in Guadeloupe by imported *Tamarixia radiata* (Hymenoptera: Eulophidae). Fruits. 2001;**56**(5):307-315

[94] Skelley LH, Hoy MA. A synchronous rearing method for the Asian citrus psyllid and its parasitoids in quarantine. Biological Control. 2004;**29**(1):14-23

[95] De León JH, Sétamou M. Molecular evidence suggests that populations of the Asian citrus psyllid parasitoid *Tamarixia radiata* (Hymenoptera: Eulophidae) from Texas, Florida, and Mexico represent a single species. Annals of the Entomological Society of America. 2010;**103**(1):100-110

[96] Michaud J. Natural mortality of Asian citrus psyllid (Homoptera: Psyllidae) in Central Florida. Biological Control. 2004;**29**(2):260-269

[97] Berg MA, Deacon VE, Fourie CJ, Anderson SH. Predators of the citrus psylla, *Trioza erytreae* (Hemiptera: Triozidae), in the Lowveld and Rustenburg areas of Transvaal. Phytophylactica. 1987;**19**:285-289

[98] Michaud J. Numerical response of Olla v-nigrum (Coleoptera: Coccinellidae) to infestations of Asian citrus psyllid (Hemiptera: Psyllidae) in Florida. Florida Entomologist. 2001;**84**(4):608-612

[99] Saldarriaga Ausique JJ, D'Alessandro CP, Conceschi MR, et al. Efficacy of entomopathogenic fungi against adult *Diaphorina citri* from laboratory to field applications. Journal of Pest Science. 2017;**90**:947-960

[100] Meyer JM, Hoy MA, Boucias DG. Morphological and molecular characterization of a Hirsutella species infecting the Asian citrus psyllid, *Diaphorina citri* Kuwayama (Hemiptera: Psyllidae), Florida. Journal of Invertebrate Pathology. 2007;**95**(2):101-109

[101] Subandiyah S, Nikoh N, Sato H, et al. Isolation and characterization of two entomopathogenic fungi attacking *Diaphorina citri* (Homoptera, Psylloidea) in Indonesia. Mycoscience. 2000;**41**:509-513

[102] Xu J, Xu X, Shakeel M, et al. The entomopathogenic fungi *Isaria fumosorosea* plays a vital role in suppressing the immune system of *Plutella xylostella*: RNA-Seq and DGE analysis of immunity-related genes. Frontiers in Microbiology. 2017;**8**:1421

[103] Casique-Valdes R, Reyes-Martinez AY, Sanchez-Peña SR, Bidochka MJ, Lopez-Arroyo JI. Pathogenicity of *Hirsutella citriformis* (Ascomycota: Cordycipitaceae) to *Diaphorina citri* (Hemiptera: Psyllidae) and *Bactericera cockerelli* (Hemiptera: Triozidae). Florida Entomologist. 2011;**94**(3):703-705

[104] Hall DG, Hentz MG, Meyer JM, et al. Observations on the entomopathogenic fungus *Hirsutella citriformis* attacking adult *Diaphorina citri* (Hemiptera: Psyllidae) in a managed citrus grove. BioControl. 2012;**57**:663-675

[105] Yang YP, Huang MD, Beattie GAC, Xia YL, Yang GC, Xiong JJ. Distribution, biology, ecology and control of the psyllid *Diaphorina citri* Kuwayama, a major pest of citrus: A status report for China. International Journal of Pest Management. 2006;**52**(4):343-352

[106] Lezama-Gutiérrez R, Molina-Ochoa J, Chávez-Flores O, et al. Use of the entomopathogenic fungi

Metarhizium anisopliae, *Cordyceps bassiana* and *Isaria fumosorosea* to control *Diaphorina citri* (Hemiptera: Psyllidae) in Persian lime under field conditions. International Journal of Tropical Insect Science. 2012;**32**(1):39-44

[107] Huvenne H, Smagghe G. Mechanisms of dsRNA uptake in insects and potential of RNAi for pest control: A review. Journal of Insect Physiology. 2010;**56**(3):227-235

[108] Huang G, Allen R, Davis EL, Baum TJ, Hussey RS. Engineering broad root-knot resistance in transgenic plants by RNAi silencing of a conserved and essential root-knot nematode parasitism gene. Proceedings of the National Academy of Sciences of the United States of America. 2006;**103**(39):14302-14306

[109] Baum J, Bogaert T, Clinton W, et al. Control of coleopteran insect pests through RNA interference. Nature Biotechnology. 2007;**25**:1322-1326

[110] Hajeri S, Killiny N, El-Mohtar C, Dawson WO, Gowda S. Citrus tristeza virus-based RNAi in citrus plants induces gene silencing in Diaphorina citri, a phloem-sap sucking insect vector of citrus greening disease (Huanglongbing). Journal of Biotechnology. 2014;**176**:42-49

[111] Taning CN, Andrade EC, Hunter WB, Christiaens O, Smagghe G. Asian Citrus psyllid RNAi pathway - RNAi evidence. Scientific Reports. 2016;**6**:38082

[112] Kumar V, Avery PB, Ahmed J, Cave RD, McKenzie CL, Osborne LS. Compatibility and efficacy of *Isaria fumosorosea* with horticultural oils for mitigation of the Asian Citrus Psyllid, *Diaphorina citri* (Hemiptera: Liviidae). Insects. 2017;**8**(4):119

[113] Zhang Q. Tests of petrol oil to control Citrus pests. China Citrus. 1985;**14**:33-34

[114] Beattie G et al. Impacts of spray volume and horticultural mineral oil concentration on control of pink wax scale and red scale in citrus orchards. Spray Oils Beyond. 2000;**58**:582-591

[115] Rae DJ, Watson DM, Huang MD, Cen YJ, Wang BZ, Beattie GAC, et al. Efficacy and phytotoxicity of multiple petroleum oil sprays on sweet orange (Citrus sinensis (L.)) and pummelo (C. grandis (L.)) in Southern China. International Journal of Pest Management. 2000;**46**(2):125-140

[116] Avery PB, Pick DA, Aristizábal LF, Kerrigan J, Powell CA, Rogers ME, et al. Compatibility of *Isaria fumosorosea* (Hypocreales: Cordycipitaceae) Blastospores with agricultural chemicals used for Management of the Asian Citrus Psyllid, *Diaphorina citri* (Hemiptera: Liviidae). Insects. 2013;**4**(4):694-711

[117] Van Vuuren S, Moll J, Da Graca J. Preliminary report on extended treatment of citrus greening with tetracycline hydrochloride by trunk injection [bacterium-like organisms]. Plant Disease Reporter (USA). 1977;**61**(5):358-359

[118] Tansey JA, Jones MM, Vanaclocha P, Robertson J. Costs and benefits of frequent low-volume applications of horticultural mineral oil for management of Asian citrus psyllid, *Diaphorina citri* Kuwayama (Hemiptera: Psyllidae). Crop Protection. 2015;**76**:59-67

[119] Qureshi JA, Kostyk BC, Stansly PA. Insecticidal suppression of Asian citrus psyllid *Diaphorina citri* (Hemiptera: Liviidae) vector of

huanglongbing pathogens. PLoS One. 2014;**9**(12):e112331

[120] Stockwell VO, Duffy B. Use of antibiotics in plant agriculture. Revue Scientifique et Technique. 2012;**31**(1):199-210

[121] McManus PS, Stockwell VO, Sundin GW, Jones AL. Antibiotic use in plant agriculture. Annual Review of Phytopathology. 2002;**40**(1):443-465

[122] Aubert B, Bové J. Effect of penicillin or tetracycline injections of citrus trees affected by greening disease under field conditions in Reunion Island. International Organization of Citrus Virologists Conference Proceedings (1957-2010). 1980;**8**(8)

[123] Bové J et al. Penicillin and tetracycline treatments of greening disease-affected citrus plants in the glasshouse, and the bacterial nature of the procaryote associated with greening. International Organization of Citrus Virologists Conference Proceedings (1957-2010). 1980;**8**(8)

[124] Nariani T, Raychaudhuri S, Viswanath S. Response of greening pathogen of citrus to certain tetracycline antibiotics. Current Science. 1971

[125] Buitendag C, Van Broembsen L. Living with citrus greening in South Africa. International Organization of Citrus Virologists Conference Proceedings (1957-2010). 1993;**12**(12)

[126] Zhang M, Powell CA, Zhou L, He Z, Stover E, Duan Y. Chemical compounds effective against the citrus Huanglongbing bacterium 'Candidatus Liberibacter asiaticus' in planta. Phytopathology. 2011;**101**(9):1097-1103

[127] Zhang M, Powell CA, Yang CY, Duan YP. Characterization of the microbial community structure in Candidatus Liberibacter asiaticus-infected citrus plants treated with antibiotics in the field. BMC Microbiology. 2013;**13**(1):112

[128] Zhang M, Powell CA, Guo Y, Doud MS, Duan Y. A graft-based chemotherapy method for screening effective molecules and rescuing huanglongbing-affected citrus plants. Phytopathology. 2012;**102**(6):567-574

[129] Abdullah TL, Shokrollah H, Sijam K, Abdullah SNA. Control of Huanglongbing (HLB) disease with reference to its occurrence in Malaysia. African Journal of Biotechnology. 2009;**8**(17):4007-4015

[130] Schwarz R, Van Vuuren S. Decrease in fruit greening of sweet orange by trunk injection of tetracyclines. Plant Disease Reporter. 1971;**55**:747-750

[131] McManus P, Jones A. Epidemiology and genetic analysis of streptomycin-resistant *Erwinia amylovora* from Michigan and evaluation of oxytetracycline for control. Phytopathology (USA). 1994;**84**:627-633

[132] Vincent C, Pierre M, Li J, Wang N. Implications of heat treatment and systemic delivery of foliar-applied Oxytetracycline on Citrus physiological management and therapy delivery. Frontiers in Plant Science. 2019;**10**:41

[133] Grondeau C, Samson R, Sands D. A review of thermotherapy to free plant materials from pathogens, especially seeds from bacteria. Critical Reviews in Plant Sciences. 1994;**13**(1):57-75

[134] Hoffman MT, Doud MS, Williams L, Zhang MQ, Ding F, Stover E, et al. Heat treatment eliminates '*Candidatus* Liberibacter asiaticus' from infected

citrus trees under controlled conditions. Phytopathology. 2013;**103**(1):15-22

[135] Lopes SA, Frare GF, Bertolini E, Cambra M, Fernandes NG, Ayres AJ, et al. Liberibacters associated with Citrus Huanglongbing in Brazil: '*Candidatus* Liberibacter asiaticus' is heat tolerant, '*Ca*. L. americanus' is heat sensitive. Plant Disease. 2009;**93**(3):257-262

[136] Su H. Modified technique of citrus shoot-tip grafting and rapid propagation method to obtain citrus budwoods free of citrus viruses and likubin organism. Proceeding of International Society Citriculture. 1984;**2**:332-334

[137] Kung-hsiang L, Ruyu Z. A preliminary study on the resistance of yellow shoot virus and citrus budwood tissue to heat. Acta Phytopathologica Sinica. 1964;**7**(1):169-175

[138] Nariani T, Bhagabati K. Studies on the therapeutic and insect vector control of the greening disease of citrus in India. International Organization of Citrus Virologists Conference Proceedings (1957-2010). 1980;**8**(8)

[139] Huang CH. Effect of hot-air treatment on likubin. Tristeza virus exocortis viroid disease of citrus. Journal of Agricultural Research of China. 1978;**2**:193-198

[140] Schwarz R, Green G. Heat requirements for symptom suppression and inactivation of the greening pathogen. International Organization of Citrus Virologists Conference Proceedings (1957-2010). 1972;**5**(5)

[141] Ehsani R, De Corcuera JIR, Khot L. The potential of thermotherapy in combatting HLB. Resource Magazine. 2013;**20**(6):18-19

[142] Yang C, Powell CA, Duan Y, Shatters R, Fang J, Zhang M. Deciphering the bacterial microbiome in Huanglongbing-affected Citrus treated with thermotherapy and Sulfonamide antibiotics. PLoS One. 2016;**11**(5):e0155472

[143] Yang CY, Powell CA, Duan YP, Shatters RG, Lin YJ, Zhang MQ. Mitigating citrus huanglongbing via effective application of antimicrobial compounds and thermotherapy. Crop Protection. 2016;**84**:150-158

[144] Fan GC, Xia YL, Lin XJ, Hu HQ, Wang XD, Ruan CQ, et al. Evaluation of thermotherapy against Huanglongbing (citrus greening) in the greenhouse. Journal of Integrative Agriculture. 2016;**15**(1):111-119

[145] Doud M, Wang Y, Hoffman M, et al. Solar thermotherapy reduces the titer of Candidatus Liberibacter asiaticus and enhances canopy growth by altering gene expression profiles in HLB-affected citrus plants. Horticulture Research. 2017;**4**:17054

[146] Jia ZC, Reza E, Zheng JQ, Xu LY, Zhou HP, Ding R. Heating characteristics and field control effect of rapid citrus huanglongbing steam heat treatment. Transactions of the Chinese Society of Agricultural Engineering. 2017;**33**(11):219-225

[147] Abdulridha J et al. Mobile thermotherapy system for treating HLB-infected citrus trees utilizing hot water and steam. In: 2018 ASABE Annual International Meeting. American Society of Agricultural and Biological Engineers; Michigan, USA. 2018

[148] Lopes SA, Luiz FQBF, Martins EC, Fassini CG, Sousa MC, Barbosa JC, et al. '*Candidatus* Liberibacter asiaticus' Titers in Citrus and acquisition rates by *Diaphorina citri* are decreased by

higher temperature. Plant Disease. 2013;**97**(12):1563-1570

[149] Aćimović SG, VanWoerkom AH, Reeb PD, Vandervoort C, Garavaglia T, Cregg BM, et al. Spatial and temporal distribution of trunk-injected imidacloprid in apple tree canopies. Pest Management Science. 2014;**70**(11):1751-1760

[150] Hu J, Jiang J, Wang N. Control of citrus huanglongbing via trunk injection of plant defense activators and antibiotics. Phytopathology. 2017;**108**(2):186-195

[151] Li J, Trivedi P, Wang N. Field evaluation of plant defense inducers for the control of citrus huanglongbing. Phytopathology. 2015;**106**(1):37-46

[152] Hu J, Wang N. Evaluation of the spatiotemporal dynamics of oxytetracycline and its control effect against citrus Huanglongbing via trunk injection. Phytopathology. 2016;**106**(12):1495-1503

[153] Planas-Riverola A, Gupta A, Betegón-Putze I, Bosch N, Ibañes M, Caño-Delgado AI. Brassinosteroid signaling in plant development and adaptation to stress. Development. 2019;**146**(5):151894

[154] Wang ZY. Brassinosteroids modulate plant immunity at multiple levels. Proceedings of the National Academy of Sciences. 2012;**109**(1):7-8

[155] Canales E, Coll Y, Hernández I, et al. 'Candidatus Liberibacter asiaticus', causal agent of Citrus Huanglongbing, is reduced by treatment with Brassinosteroids. PLoS One. 2016;**11**(1):e0146223

[156] Stock D, Holloway PJ. Possible mechanisms for surfactant-induced foliar uptake of agrochemicals. Pesticide Science. 1993;**38**(2-3):165-177

[157] Castro J, Ojeda C, Cirelli AF. Advances in surfactants for agrochemicals. Environmental Chemistry Letters. 2014;**12**(1):85-95

[158] Gurpreet K, Singh S. Review of nanoemulsion formulation and characterization techniques. Indian Journal of Pharmaceutical Sciences. 2018;**80**(5):781-789

[159] Bouchemal K, Briançon S, Perrier E, Fessi H. Nano-emulsion formulation using spontaneous emulsification: Solvent, oil and surfactant optimisation. International Journal of Pharmaceutics. 2004;**280**(1-2):241-251

[160] Yang C, Powell CA, Duan Y, Shatters R, Zhang M. Antimicrobial Nanoemulsion formulation with improved penetration of foliar spray through Citrus leaf cuticles to control Citrus Huanglongbing. PLoS One. 2015;**10**(7):e0133826

[161] Yang CY, Powell CA, Duan YP, Zhang MQ. Characterization and antibacterial activity of oil-in-water nano-emulsion formulation against Candidatus Liberibacter asiaticus. Plant Disease. 2016;**100**(12):2448-2454

[162] Singh S, Singh BK, Yadav SM, Gupta AK. Applications of nanotechnology in agricultural and their role in disease management. Research Journal of Nanoscience and Nanotechnology. 2015;**5**:1-5

[163] Douglas AE. Strategies for enhanced crop resistance to insect pests. Annual Review of Plant Biology. 2018;**69**:637-660

[164] Viloria Z, Drouillard DL, Graham JH, Grosser JW. Screening triploid hybrids of 'Lakeland' Limequat

for resistance to Citrus canker. Plant Disease. 2004;**88**(10):1056-1060

[165] Zhang X, Francis MI, Dawson WO, et al. Over-expression of the Arabidopsis NPR1 gene in citrus increases resistance to citrus canker. European Journal of Plant Pathology. 2010;**128**:91-100

[166] Durrant WE, Dong X. Systemic acquired resistance. Annual Review of Phytopathology. 2004;**42**:185-209

[167] Backer R, Naidoo S, van den Berg N. The NONEXPRESSOR OF PATHOGENESIS-RELATED GENES 1 (NPR1) and related family: Mechanistic insights in plant disease resistance. Frontiers in Plant Science. 2019;**10**:102

[168] Cao H, Bowling SA, Gordon AS, Dong X. Characterization of an Arabidopsis mutant that is nonresponsive to inducers of systemic acquired resistance. The Plant Cell. 1994;**6**(11):1583-1592

[169] Cao H, Li X, Dong X. Generation of broad-spectrum disease resistance by overexpression of an essential regulatory gene in systemic acquired resistance. Proceedings of the National Academy of Sciences. 1998;**95**(11):6531-6536

[170] Dutt M, Barthe G, Irey M, Grosser J. Transgenic Citrus expressing an Arabidopsis NPR1 gene exhibit enhanced resistance against Huanglongbing (HLB; Citrus greening). PLoS One. 2015;**10**(9):e0137134

[171] Wang Y, Zhou L, Yu X, Stover E, Luo F, Duan Y. Transcriptome profiling of Huanglongbing (HLB) tolerant and susceptible Citrus plants reveals the role of basal resistance in HLB tolerance. Frontiers in Plant Science. 2016;**7**:933

[172] Pulizzi F. Nano in the future of crops. Nature Nanotechnology. 2019;**14**:507

[173] Elmer W, White JC. The future of nanotechnology in plant pathology. Annual Review of Phytopathology. 2018;**56**:111-133

Chapter 2

Anthracnose Disease of Mango: Epidemiology, Impact and Management Options

Frederick Kankam, Stephen Larbi-Koranteng, Joseph Adomako, Joseph Kwowura Kwodaga, Isaac Boatey Akpatsu, Yaw Danso and Elias Nortaa Kunedeb Sowley

Abstract

Mango is one of the frequently cultivated seasonal fruit crops in several tropical and subtropical regions. It is consumed as whole fruits apart from serving as raw materials for most industries that are into mineral production. Mango production is, however, constrained by diseases, pests, and poor post-harvest handling of fruits. Anthracnose disease, caused by *Colletotrichum gloeosporioides* Penz and Sacc, is one of the most important yields limiting constraint in mango production across the globe. The disease occurs in both the field and post-harvesting. In the field, it affects aboveground parts, such as the stem, branches, leaves, flowers, and fruits. Anthracnose disease reduces the shelve life and marketability of mango fruit. In Ghana, anthracnose disease is responsible for about 30% yield/fruit loss. Most farmers do not control it, although some have resorted to the application of various fungicides not registered for mango anthracnose disease management in Ghana. This chapter will highlight on the importance of the disease on the mango industry in Ghana, control strategies currently employed thereby reducing the over-reliance on chemical control option and propose ways to minimize the effect of the disease in the country.

Keywords: aetiology, anthracnose, botanicals, *Colletotrichum* spp., epidemiology, mango

1. Introduction

Mango (*Mangifera indica* L) is a major fruit crop with an immense economical relevance in Ghana. Due to the special qualities of its fruits, it is also considered as one of the most frequently considered fruits in various fruit markets across the global. Some of the qualities mango are delicious taste, beautiful colour, excellent carotene content, excellent flavour and attractive flavour [1]. The crop is known to contribute up to a half of the overall tropical fruits that are produced globally [2]. Africa contributes about 10–15% of the total mango produced per annum. In Ghana and other neighbouring countries, the crop serves as one of the most frequent non-traditional fruits exported to other countries. Mango in Ghana is currently in high demand in the form of both fresh and dry fruits, juice and flavours, and jams. In as much as mango production in Ghana and other production countries has seen some increase in recent times, it is frequently faced with challenges such as pests and diseases attack. One of the most devastating diseases is the anthracnose disease. The mango anthracnose disease caused by *Colletotrichum gloeosporioides* Penz and Sacc is a major fungal disease of economic importance to mango production across the globe [3–6]. To manage this menace, various strategies has been adopted by various stakeholders in the mango production value chain [7]. These strategies ranges from pre-harvest to post-harvest. Both the pre-harvest and post-harvest control measures are successfully implemented through proper field hygiene, after harvest treatments or a combination of the two. However, the management strategy chosen at any point by an individual must be consumer friendly, economical, and environmentally friendly. In composite, these management strategies can further be categorised into cultural, chemical, biological, use of resistant cultivars, and/or their integration [7].

2. Impact of anthracnose on mango production in Ghana

Different kinds of diseases affect productivity of several plants contributing significantly to global food/fruit shortage. Destructions/Catastrophe caused by the outbreak of plant diseases in multiple plants as well as its impact on humans such as starvation, malnutrition and death have been well publicized. In Ghana, the impact of diseases on tree and horticultural crops have been reported and well documented. Mango production in Ghana has gained prominence in recent times with the projection of overtaking cocoa as the most economically valuable crop. The potential gains due the country from mango production is however, threatened by the existence and outbreak of multiple diseases due to the crop's susceptibility to several fungal pathogens. Mango is highly susceptible to diseases due to the high water-nutrient content which serves as a perfect media for fungal pathogen development [8]. Production and export of mango in Ghana keeps dwindling due to the prevalence and severity of diseases such as the anthracnose [9]. Among the many diseases limiting mango production is the Anthracnose disease caused by the fungus *Colletotrichum gloeosporioides* Penz. & Sacc. It is a major limitation to mango production as it has been reported to has led to a yield loss of 30% in the Yilo Kobo District of Ghana [10]. Similarly, about 39% yield loss in mango production in India has been attributed to anthracnose disease [11]. The disease is so devastating in crop production that it has also been reported that 60% of harvested avocado fruits affected by the disease were discarded and rendered unmarketable in Kenya [12].

Although the disease leads to massive losses in the field, the most significant loss occurs during post-harvest conditions. The post-harvest damage due to the disease is of more interest and economically relevant because it reduces the fruits quality as well as shelf life which invariably affects standard for the export market. This pose great challenge to various actors along the value chain especially those involved in international trade as low-quality fruits cannot be marketed on the international market. Although in some instances, producers and traders may sell these low-quality fruits on the local market, differences in revenue from export and local markets leads to serious economic loss to all stakeholders as well as mango exporting countries like Ghana as they are denied of foreign exchanges.

Attempts to mitigate the effects of anthracnose disease on mango production have centred mainly on the employment of conventional pesticides to treat trees and even fruits at harvest and storage. Although effective in minimizing the severity of the disease, this strategy is limited due to the growing negative concerns associated with fungicides on the environment, human health and rapid surge in antifungal resistance by several pathogens. In addition to these, is the high cost of these pesticides leading to increased cost of production to the farmer and reduced income which can further increases poverty in the rural economy.

3. Aetiology and epidemiology of anthracnose disease of mango

3.1 Aetiology of anthracnose disease of mango

Anthracnose disease of mango is one of the serious diseases of mango worldwide as it affects the crop from nursery through growth (pre-harvest) to harvested fruits (post-harvest) [11, 13, 14]. Many reporters have confirmed that two closely related species of Colletotrichum are involve in disease initiation and development. These species are *Colletotrichum gloeosporioides* Penz and Sacc. (Teleomorph: *Glomerella cingulata*) and *Colletotrichum acutatum* (Teleomorph: *Glomerella acutata*) [15–17]. In Ghana, literature has predominantly reported that the *C. gloeosporioides* is the causal agent of the disease [18, 19].

3.2 Epidemiology and disease cycle

The disease initiation, growth and development of the pathogen are influenced greatly by the nutritional and environmental factors of the host plant (Mango). Among the nutritional factors are sources of nitrogen and carbon [20] whereas the environmental factors for disease development wet, humid, hot weather [21] as well as pH and Temperature [20]. The pathogen is also considered generally inactive in dry weather [3].

It is reported that the source of primary inocula of the disease if the abundance of conidia in the tree canopy [3]. The disease attacks every part of the plant i.e. young nursery seedling leave and twigs, young flowers/flower clusters (panicles), young fruits as well as mature fruits. The cycle of the disease begins with the dissemination of the conidia (asexual spore) of the pathogen which are dispersed passively through rain splashes or irrigation water. The plant is inoculated when conidia land on the infection courts or sites (most of which are above ground parts of the plant). Infection and disease development are also achieved on young and immature fruit and tissues, conidia are germinated and penetrated in the cuticles and epidermis to ramify through the tissues. On mature fruits, infections occur immediately when spores penetrate the cuticle. However, they remain at a latent state until climacteric fruit ripening begins. During disease symptoms development, black, sunken, rapidly expanding lesions develop on affected organs. On leaves, symptoms are seen as small, brown-to-black lesions on most affected parts.

On fruits where major destruction occur both in pre and post-harvest, symptoms of black/dark is visibly as well as slightly sunken lesion with irregular shape, then gradually enlarges and subsequently causes fruit rot. The first appearance of symptoms as spots usually coalesces and penetrate deep into the fruit, resulting in extensive fruit rotting. In most green fruits, infections remain latent and largely invisible until ripening, that fruit that appear healthy at harvest can develop significant symptoms upon ripening. The second symptom type of the fruit is one that commonly refers to as an "alligator skin" in which on the fruits consist of a "tear stain" symptoms, in which appear a linear necrotic region on the fruits that experience superficial cracking of the epidermis. Lesions on stem and fruits may produce visible, pinkish-orange spore masses under wet conditions. During reproduction of the pathogen, sticky masses of conidia and production of fruiting bodies (acevulli) on symptomatic tissues during humid conditions. Repeated cycles of the disease may occur as a result of continuous multiplication of the fungus during the season which can serious epidemics. In the absence of favourable conditions, the pathogen overwinters in plant debris such as defoliated leaves, dead branches and other volunteer tree species.

4. Management strategies of mango anthracnose disease

4.1 Cultural control

One of the main conditions that favours the development of the mango anthracnose is high humidity. It is therefore important to establish mango orchards in areas with a well-defined dry season. This creates an unfavourable condition for the development of the disease [12]. In most tropical regions, flowering in mango plants occurs usually in the dry seasons. Nevertheless, this is dependent on other factors such as photoperiodism, growth stage of the shoot, specie of the mango, ambient temperature, as well as the nutritional state of the soil under cultivation. This brings some level of variations in the flowering period of mango from region to region. It has been noted that a very early flowering, usually before the dry season leads to a huge infection of the disease due to the higher relative humidity at the time of fruit formation and establishment [22, 23]. From another standpoint, a worst scenario of infection is seen when mango trees flower at the latter part of the dry season or when early-stage fruit development coincides with the peak of the raining season [24]. This can lead to a disease incidence of up to 90% in such occasions. For proper anthracnose disease management in the tropics, a plausible approach must be adopted to avoid situations where flowering an early-stage fruit development will coincide with the rainy season. A proper adherence to this approach could cumulatively, leads to a disease incidence and severity near zero present. However, despite the effectiveness of this strategy, it is near inapplicable in the subtropical regions where the main stimulants for flowering in mango is atmospheric temperature rather than lower water deficit. One of the most effective ways of managing the incidence and severity of this disease in these areas is the foliar application of a growth and flowering retardant (e.g. Paclobutrazole) either singly or in combination with potassium nitrate [24, 25]. By that, flowering in mango can be advanced by several days or weeks. Furthermore, traditional cultural practices such as the removal and proper incineration of infected plant parts such as fruits, branches, leaves among others can also be reliable in the cultural management system of mango anthracnose disease. In addition, practices such as removal of dry, infected, or malformed panicles and fruits can serve as one of the effective ways of managing the disease. However, these later practices can be laborious and time consuming on large area farms. Research has also revealed that wrapping of young developing fruits in paper bags as a way of creating barriers between the fruits and any possible inoculum of the disease can immensely reduce the incidence and severity of the disease. However, his method can also reduce the bright red/yellow colour observed in ripped mangoes. This, on a broader perspective could insinuate that choosing a particular method or practice in managing the disease could highly depend on the objective and the target market of the farmer. For example, most industries that would rather prefer chemical free fruits to bright coloured ones will rather appreciate this approach as compared to the application of various synthetic fungicides other methods such as chemicals.

4.2 Resistant varieties

The use of cultivars that are resistant to disease has always been one of the most effective and economical ways of managing diseases in many crops, including mango. However, in mango, several studies have shown that most cultivars that are commercially available to farmers are susceptible to the disease. Nevertheless, some experiments in controlled environments have reported that there are some few genotypes with some considerable levels of resistance to the mango anthracnose disease (**Table 1**). These traits can be harnessed in future breeding intervention to development some potential cultivars with higher resistance to the disease [26].

4.3 Chemical control

Just like in most scenarios, the application of chemicals seems to be one of the most effective and fastest ways of managing the mango anthracnose disease. Nevertheless,

Name	Designation	Country
'Carrie'	R	Australia
'Caraboa Florigon'	R	Australia
'Tommy Atkins'	R	Australia
'Saigon'	R	Australia
'Kensington Pride'	MR	Australia
'Palmer'	R	Philippines
'Siam'	R	Philippines
'Velci Colomban'	R	Philippines
'Joe Welch'	R	Philippines
'Fernandin'	MR	Philippines
'Arumanis'	MR	Philippines
'Edward'	MR	Philippines
'Gedong	MR	Philippines
Tjenkir'	MR	Philippines
Paris', 'Fairchild	R	Hawai'i
'Rapoza'	R	Hawai'i
'Haden'	MR	Hawai'i
'Zill'	R	Florida

R = resistant; and MR = moderately resistant.

Table 1.
Mango cultivars resistant to the anthracnose disease [21].

under severe pressures, one has to conduct up to 25 applications in a season to achieve desired results. Also, the choice of fungicide is also dependent on the variation in requirements of proposed destination or purpose of the fruit. For example, Dithiocarbamate (an effective active ingredient for anthracnose disease control) produces ethylene thiourea (ETU) as a bio-product. Due to this, fruits that has a history of the application of ethylene-bisdithiocar- bamates such as mancozeb and maneb are prohibited in the United States. Most frequently, copper fungicides are recommended for the control of most fungal disease, which the mango anthracnose is not an exception. In as much as some success stories has been documented, there are also contrasting studies that has revealed that under higher disease pressures, copper-based fungicides are usually less effective as compared to the carbamates. Not only that, few reports have also reported some phytotoxicity associated with the application of fungicides of various sorts. Post-infection fungicides such as the benzimidazoles has also been used in combination with protectant fungicides such as the imidazole prochloraz to retard the build-up pressure of the mango anthracnose disease [27–29]. This chemical s usually affects the mycelia development of the fungi by supressing the synthesisation of ergosterols, a very important component of the cell membrane of a fungus [30]. One advantage for this combination is that it also retards the build-up of resistance in the population of the pathogen. Other reports have also documented the use of clarified hydrophobic neem oil (70%) to effectively control the mango anthracnose disease.

4.4 Biological control

Biological approach of managing disease is the use of antagonistic organisms in managing diseases. In the cases of mango anthracnose, an array of microorganisms have been postulated to have an antagonistic association on its causative agent (*C. gloeosporioides*) [31]. However, even though this looks very promising, there is no current documentation of commercial implementation of this finding. In expense, keen attention has rather been placed on the postharvest control method. Perhaps, this could be as a result of the ability to manage controlled environment conditions as compare to field situations. Nevertheless, the antagonistic potentials from such biological agents can be harnessed as an intervention to developing a more robust management approach to the disease. An example was cited in a study on the efficacy of essential oils in managing the mango anthracnose disease [32]. As all other disease management approaches, the efficacy of biological control method can be enhanced with an integration of other control measures [33].

4.5 Current research on the management of anthracnose disease in mango with botanicals

Anthracnose an important pre and post-harvest disease of mango infests mango parts such as leaves, twigs, flowers and fruits. The disease can dwindle the productivity of infected mango plants as well as the quality of mango fruits; hence, resulting in economic losses. There is therefore the need to manage mango anthracnose disease in order to enhance the health and productivity of mango plants and also maintain the quality of mango fruits. Several methods such as culture, use of synthetic fungicides, resistant varieties and biological have been employed in the management of mango anthracnose diseases. The use of synthetic fungicides is one of the major methods employed in managing mango anthracnose disease. However, the use of synthetic fungicides is usually accompanied with challenges such as pollution of the environmental, development of pathogenic resistance, residual toxicity and harmful effects on humans [34, 35]. The excessive and improper use of synthetic fungicides can result in the accumulation of fungicidal residue in plants and plant organs used as food which may pose health risk to consumers. For instance, the residual effects of some synthetic pesticides have been recorded in fruits of melon, guava, orange, peach and mango at levels toxic to human consumption [36]. There is therefore the need to find environmentally friendly alternatives in managing anthracnose disease of mango.

The use of botanical fungicides in plant disease management can help promote sustainable agriculture since botanicals are natural, easily biodegraded into harmless substances; hence do not persist in the environment and plant parts used as food. The antifungal activities of botanicals result from the phytochemical they contain. These phytochemicals which are secondary metabolites produced by plants in nature play an important role in the ability of plant to defend themselves against phytopathogens. Phytochemicals have antimicrobial properties [37]; hence, can prevent or reduce infection when applied on plants or plant parts.

Currently, several studies have documented the use of botanical in the management of anthracnose of mango [38–40]. At the pre-harvest of mango anthracnose disease, studies has showed that aqueous extracts leave of *Eucalyptus camaldulensis* and *Azadirachta indica* inhibited the mycelia growth of *C. gloeosporioides in vitro* and also under field conditions, the foliar application of the extracts reduced the incidence and severity of anthracnose on mango plants [38]. Furthermore, literature has also reported that mango fruits treated by dipping in aqueous extracts of *Ruta chalepensis* at concentration of 50 grams of the powdered plant material in 100 ml of distilled water before storage remarkably reduced the occurrence of anthracnose disease, maintained the quality and marketability of the fruits [41]. Some botanicals in comparison to synthetic fungicides have exhibited an equivalent level of antifungal activity against *Colletotrichum* spp. For instance, studies has showed that essential oil of basil leaves inhibited the mycelia growth of *C. acutatum* that caused anthracnose disease in fruits of mango cat hoc variety, and also significantly reduced the incidence and severity of anthracnose on the mango fruits comparable to those treated with the synthetic fungicide Tolent 50 WP (Prochloraz) [40].

Botanicals can be used for the eco-friendly management of both the pre- and post-harvest anthracnose disease of mango [39]. Their studies reported that the pre-harvest anthracnose disease of mango was effectively controlled by the foliar application of aqueous extracts of *Azadirachta indica, Eucalyptus camaldulensis, Allium sativum, Zingiber ofcinale* and *Calotropis procera,* and at the post-harvest stage, mango fruits sprayed with the botanicals reduced the anthracnose infection of the fruits and improved their quality. Although many botanicals have exhibited some potential to be used in managing anthracnose disease of mango, the main challenge is that most of these findings have not moved beyond the research stage and as such not readily available to mango growers and marketers. To encourage the use of botanical in the management of mango anthracnose disease, there is the need to develop and register botanicals which have exhibited the potential to manage the disease into commercial botanical fungicides products which can easily be assessed by mango growers and marketers.

5. Postharvest treatments

Usually, fruits that are contaminated or infected with disease inoculum shortly before harvesting do not show any symptom of the disease due to latency. This state of infection at the time of harvest leads to a lethal level of the disease shortly after harvest. This leads to huge quantitative loss of the produce. As much as synthetic fungicide applications still remains as one of the most basic and effective ways of managing the disease [42], their application to the fruit after harvest seams not to be safe for human consumption as it could lead to fatality [43]. This calls for an eco-friendlier approach in managing the disease postharvest, as an effective alternative for chemical control. Various treatments are applied to the fruit to prevent or perhaps retard the development process of the disease [44]. This treatment usually has to do with temperature manipulation [45]. At the onset of ripening of the fruit postharvest, fruits can be refrigerated at 10°C to retard the development of the disease. However, it must be noted that fruits should not be chilled before ripening to reduce or avoid chilling injury and further reduce the quality of the fruit. Fruits may also be dipped in hot water (usually at temperatures 50–60°C) for a duration of up to 15 minutes [46]. The effectiveness of this method has made it to been known as one of the most effective and environmentally friendly postharvest control methods of the mango anthracnose and has been recommended by several disease control departments across the globe. Fruits may also be exposed to vapour heat, forced-air dry heat for about 3–6 hours at appropriate temperatures. However, in all these treatments, the temperature levels and duration of exposure depends greatly on the variety of the subject. Also, various treatment combinations has been found to be effective in the management of the disease [47]. For example, beromyl has been reported to be very effective in the management of quiescent infections of the anthracnose disease in mango when it is coupled with the various hot water treatments. On the other hand, prochloraz has also been found very effective when combined with cold water treatment. However, this was said to be less effective as compared to the beromyl treatment. It is also important to note that the effectiveness of any of these methods has a direct baring with the variety or cultivar. This means that in choosing a particular postharvest management method, one as to put into consideration the variety of the mango. For example, morphological feathers such as skin thickness of mango varies from variety to variety. For optimum effectiveness, treatment should be applied to varieties as may be applicable. The overall effectiveness of postharvest treatment approaches is said to be moderate in the management of the disease. This could perhaps be because treatments are only aimed predominantly at managing the disease and not as a protectant. In summary, the following treatments has been recommended to be effective for postharvest management of mango anthracnose; scrubbing with 1% NaOCl, hot water dip (50-55oC for 3–10 minutes), hot benomyl dip (500–1000 ppm), hot/cold prochloraz dip (400–l000 ppm), hot imazalil (1000 ppm), hot water +20 k RAD irradiation, hot water +75 k RAD irradiation, hot benlatc/iprodionc (1000 ppm) + 75 k RAD irradiation + waxing [48, 49].

6. Colletotrichum: current status and future directions

6.1 Colletotrichum: current status

Colletotrichum species are important cosmopolitan pathogens of many plant species. Globally, *Colletotrichum* can cause anthracnose disease on various types of suitable host plants at the pre- and post-harvest stages resulting in economic losses in crop production.

In the absence of the host plant, the inocula of *Colletotrichum* species can survive unfavourable conditions on plant debris, alternate and collateral hosts, and volunteer crops on harvested crop fields. There has been continuous first reports of anthracnose disease caused by *Colletotrichum* species on various plants across the world [50–53]. This indicates that the inocula of *Colletotrichum* species are persistent in the environment and expanding their plant host range; hence, remain a major threat to crop cultivation.

Cross infection of anthracnose disease caused by *Colletotrichum* species from one species of plant or its product to another has been reported [54, 55]. A great diversity of *Colletotrichum* species can cause anthracnose disease on a particular host plant and also cross infect another suitable host. *Colletotrichum* species such as *C. asianum, C. cliviicola, C. cordylinicola, C. endophytica, C. fructicola, C. gigasporum, C. gloeosporioides, C. karstii, C. liaoningense, C. musae, C. scovillei, C. siamense* and *C. tropicale* were found to cause mango anthracnose disease and also cross infect other crops [56].

Methods such as chemical, culture, biological and use of resistant varieties have been employed in the management of anthracnose disease. A major challenge confronting the management of anthracnose disease caused by *Colletotrichum* is how to properly identify closely related species of the pathogen causing the disease, since some of the *Colletotrichum* species have similar morphological characteristics and show similar symptoms on infected plants [56, 57]. To avoid the issue of improper identification, several studies have advocated the use of molecular methods for accurate identification of *Colletotrichum* species [54, 56, 57]. Accurate identification of Colletotrichum species causing a particular plant disease would aid in the proper management of the disease.

6.2 Colletotrichum: future directions

Anthracnose disease of plants caused by *Colletotrichum* species is real and remain a major threat to crop production now and into the future. Going forward, the general public should be sensitized about plant diseases caused by *Colletotrichum* species. This would make the general public aware of the disease and as such play a key role in managing the disease. For example, people who are conscious of the disease would help minimize the rate at which they aid in spreading the *Colletotrichum* inocula from one anthracnose infected plant or its products to another of the same species or others of different species which may also be suitable host for the pathogen.

Going forward, there is the need to properly identify the particular species of *Colletotrichum* causing a plant disease to allow for the proper management of the disease. Expects in the field of fungi identification should be consulted when identifying *Colletotrichum* species and the identified species confirmed using molecular methods. Anthracnose disease of plants caused by *Colletotrichum* species should be of public concern since it can pose a major threat to food security in the future. More studies should be conducted on the epidemiology, aetiology and sustainable methods of managing diseases caused by *Colletotrichum* species.

7. Conclusions

Mango anthracnose disease is one of the most important economic diseases of mango in many mango production areas across the globe. It affects the almost all above ground parts of the plant and demands both preharvest and postharvest approaches for effective management [17]. The management regimes range from timely application of appropriate fungicides on the field, smart use of traditional cultural practices, the use of resistant cultivars, to postharvest treatments such as dip treatments and refrigeration. Adequate knowledge of this disease is one of the most essential requisites in choosing the most appropriate control measure for an optimum control.

Conflict of interest

Authors declare no competing interest.

Author details

Frederick Kankam[1*], Stephen Larbi-Koranteng[2], Joseph Adomako[1], Joseph Kwowura Kwodaga[1], Isaac Boatey Akpatsu[1], Yaw Danso[3] and Elias Nortaa Kunedeb Sowley[1]

1 Department of Crop Science, Faculty of Agriculture, Food and Consumer Sciences, University for Development Studies, Tamale, Ghana

2 Department of Crop and Soil Science, College of Agricultural Education, Akenten Appiah-Menka University of Skills Training and Entrepreneurial Development, Asante-Mampong, Ghana

3 CSIR-Crops Research Institute, Kumasi, Ghana

*Address all correspondence to: fkankam@uds.edu.gh

References

[1] Arya A. Tropical Fruit Diseases and Pests. Ludhiana: Kalyani Publications; 2014

[2] FAO. FAO Statistics: Final 2009 Data. (FAOSTAT). 2011. Available from: http://faostat.fao.org/site/567/DesktopDefault.aspx?PageID=567#ancor

[3] Arauz LF. Mango anthracnose: Economic impact and current options for integrated management. Plant Disease. 2000;**84**(6):600-611

[4] Campbell A, Ploetz R, Gutierrez O. Using digital image analysis to evaluate disease resistance. In: The Americas Cacao Breeders' Working Group 2nd Annual Meeting, San Salvador, El Salvador; Sep 9, 2015

[5] Awa OC. First report of fruit anthracnose in mango caused by Colletotrichum gloeosporioides in South-Western Nigeria. International Journal of Scientific and Technology Research. 2012;**1**(4):20-34

[6] BBS. Yearbook of Agricultural Statistics-2017. Bangladesh: Bangladesh Bureau of Statistics, Statistics Division, Ministry of Planning; 2018

[7] Prusky D. Post-Harvest Pathology. 7th ed. London: Springer; 2014

[8] Konsue W, Dethoup T, Limtong S. Biological control of fruit rot and anthracnose of postharvest mango by antagonistic yeasts from economic crops leaves. Microorganisms. 2020;**8**(3):317. DOI: 10.3390/microorganisms8030317

[9] Asare EK, Donfeh O, Avicor SW, Pobee P, Bukari Y, Amoako-Attah I. Colletotrichum gloeosporioides causes and outbreak of Anthrachnose disease of cocoa in Ghana. Sout African Journal of Plant and Soil. 2021;**38**(2):107-115

[10] Honger O. Charachterisation of the Causal Agent of Mango Anthracnose Disease in Ghana. Ghana: University of Ghana; 2014

[11] Prakash OM, Srivastava KC. Mango Diseases and their Management. A World Review. New Delhi, India: Today and Tomorrow's Printers and Publishers; 1997

[12] Kimaru KS, Muchemi KP, Mwangi JW. Effects of anthracnose disease on avocado production in Kenya. Cogent Food and Agriculture. 2022;**1**:6. DOI: 10.1080/23311932.2020.1799531

[13] Jeffries A, Dodd JC, Jeger MJ, Plumbley RA. The biology and control of Colletotrichum species on tropical fruit crops. Plant Pathology. 1990;**39**:343-366

[14] Bally ISE, Hofman PJ, Irving DE, Coates LM, Dann EK. The effects of nitrogen on postharvest disease in mango (*Mangifera indica* L. 'Keitt'). Acta Horticulturae. 2009;**820**:365-370

[15] Freeman S, Katan T, Shabi E. Characterization of Colletotrichum species responsible for anthracnose disease of various fruits. Plant Disease. 1998;**82**:596-605

[16] Peres NAR, Kuramae EE, Dias MSC, De-Souzza NL. Identification and characterisation of Colletotrichum species affecting fruit after harvest in Brazil. Journal of Phytopathology. 2002;**105**:128-134

[17] Tarnowski TL, Ploetz R. Assessing the role of *Colletotrichum gloeosporioides* and *C. acutatum* in mango anthracnose

in South Florida. Phytopathology. 2008;**98**:S155

[18] Oduro KA. Checklist of plant pests in Ghana. In: Diseases. Vol. 1. Accra, Ghana: Plant Protection and Regulatory Service Directorate, Ministry of Food and Agriculture; 2000. p. 105

[19] Offei SK, Cornelius EW, Sakyi-Dawson O. Crop Diseases in Ghana and Their Management. Tema, Ghana: Smartline Publishing Limited; 2008

[20] Eicher R, Ludwig H. Influence of activation and germination on high pressure inactivation of ascospores of the mould Eurotium repens. Comparative Biochemistry and Physiology. 2002;**131**:595-604

[21] Nelson SC. Mango Anthracnose (*Colletotrichum gloeosporioides*). Publication PD-48. Hawaii, USA: College of Tropical Agriculture and Human Resource; 2008

[22] Dodd JC. The effect of climatic factors on Colletotrichum gloeosporioides, causal agent of mango anthracnose, in the Philippines. Plant Pathology. 1991;**40**(4):568-575

[23] Udhayakumar R, Usha RS. Epidemiological and nutritional factors on growth of Colletotrichum gloeosporioides (Penz.) Penz. And Sacc. Annals of Plant Protection Sciences. 2010;**18**(1):159-163

[24] Sanders GM, Korsten L. Comparison of cross inoculation potential of South African avocado and mango isolates of Colletotrichum gloeosporioides. Microbiological Research. 2003;**158**(2):143-150

[25] Bally IS, Hofman PJ, Irving DE, Coates LM, Dann EK. The effects of nitrogen on postharvest disease in mango (*Mangifera indica* L. 'Keitt'). In: VIII International Mango Symposium. Vol. 820. 2006. pp. 365-370

[26] Akem CN. Mango anthracnose disease: Present status and future research priorities. Plant Pathology Journal. 2006;**5**(3):266-273

[27] Everett KR, Owen SG, Cutting JGM. Testing efficacy of fungicides against postharvest pathogens of avocado (Persea americana cv. Hass). Plant Protection. 2005;**58**:89-95. DOI: 10.30843/nzpp.2005.58.4260

[28] Scheepers S, Jooste A, Alemu ZG. Quantifying the impact of phytosanitry standards with specific reference to MRLs on the trade flow of south African avocados to the EU. Agrekon. 2007;**46**(2):260-273. DOI: 10.1080/03031853.2007.9523771

[29] Smith LA, Dann BEK, Leonardi J, Dean JR, Cooke AW. Exploring on traditional products for management of postharvest anthracnose and stem end rot in avocado. In: Proceedings VII World Avocado Congress. Cairns, Australia, 5-9 September 2011

[30] Bill M, Dharini-Sivakumar A, Thompson K, Korsten L. Avocado fruit quality management during the postharvest supply chain. Food Review International. 2014;**30**(3):169-202. DOI: 10.1080/87559129.2014.907304

[31] Kefialew Y, Amare A. Postharvest biological control of anthracnose (Colletotrichum gloeosporioides) on mango (Mangifera indica). Postharvest Biology and Technology. 2008;**50**(1):8-11

[32] Chala A. Survey of mango anthracnose in southern Ethiopia and in-vitro screening of some essential oils against Colletotrichum gloeosporioides.

International Journal of Fruit Science. 2014;**14**(2):157-173

[33] Singh P. Integrated management of storage anthracnose of mango. Journal of Mycology and Plant Pathology. 2011;**41**:63

[34] Gupta PK. Toxicity of fungicides. In: Veterinary Toxicology. Cambridge, MA, USA: Academic Press, Inc; 2018. pp. 569-580

[35] Khaskheli MI. Mango Diseases: Impact of Fungicides. In: Horticultural Crops. London: IntechOpen; 2020

[36] Qamar A, Asi R, Iqbal M, Nazir A, Arif K. Survey of residual pesticides in various fresh fruit crops: A case study. Polish Journal of Environmental Studies. 2017;**26**(6):2703-2709

[37] Khameneh B, Iranshahy M, Soheili V, Bazzaz BSF. Review on plant antimicrobials: A mechanistic viewpoint. Antimicrobial Resistance and Infection Control. 2019;**8**(1):1-28

[38] Kumari P, Singh R, Punia R. Evaluation of fungicides and botanicals against mango (*Mangifera indica*) anthracnose. Current Journal of Applied Science and Technology. 2017;**23**(3):1-6. DOI: 10.9734/CJAST/2017/35899

[39] Iram S, Laraib H, Ahmad KS, Jaffri SB. Sustainable management of Mangifera indica pre-and post-harvest diseases mediated by botanical extracts via foliar and fruit application. Journal of Plant Diseases and Protection. 2019;**126**(4):367-372

[40] Danh LT et al. Use of essential oils for the control of anthracnose disease caused by Colletotrichum acutatum on post-harvest mangoes of cat Hoa Loc variety. Membranes (Basel). 2021;**11**(9):719

[41] Alemu K, Ayalew A, Woldetsadik K. Antifungal activity of plant extracts and their applicability in extending the shelf life of mango fruits. Food Science and Quality Management. 2014;**33**:47-53

[42] Castro JV. Effect of post-harvest application of fungicides on control of anthracnose in mangoes. Boletim do Institute de Technologica de Alimentoes. 1985;**21**:447-451

[43] Pavitra Kumari R, Singh R. Anthracnose of mango incited by *Colletotrichum gloeosporioides*: A comprehensive review. International Journal of Pure Applied Bioscience. 2017;**5**(1):48-56

[44] Chaudhari MA. Management of Post-Harvest Anthracnose of Mango (Mangifera indica L.). Navsari, Gujarat, India: Navsari Agricultural University; 2013

[45] Lai AA, Simon S. Post-harvest management of anthracnose rot of mango (Mangifera indica L.). Annals of Plant protection Sciences. 2013;**21**(1):121-124

[46] Mirshekari A. Effect of hot water dip treatment on postharvest anthracnose of banana var. Berangan. African Journal of Agricultural Research. 2012;**7**(1):6-10

[47] Chowdhury MNA, Rahim MA. Integrated crop management to control anthracnose (Colletotrichum gloeosporioides) of mango. Journal of Agriculture and Rural Development. 2009;**7**(1-2):115-120

[48] Rajwana IA. Effect of combined application of fungicides and hot water quarantine treatment on postharvest diseases and quality of mango fruit. Pakistan Journal of Botony. 2011;**43**(1):65-73

[49] Sepiah M. Effectiveness of hot water, hot benomyl and cooling on postharvest diseases of mango cv. Harumani. ASEAN Food Journal. 1986;**2**:117-120

[50] Oliveira SA, Silva LL, Diamantino MS, Ferreira CF. First report of *Colletotrichum theobromicola* and *C. siamense* causing anthracnose on cultivated and wild cassava species in Brazil. Plant Disease. 2018;**102**(4):819

[51] Li Q et al. Colletotrichum species associated with mango in southern China. Scientific Reports. 2019;**9**(1): 1-10

[52] Ben H et al. First report of Colletotrichum capsici causing anthracnose on Alocasia macrorrhizos in China. Plant Disease. 2021;**105**(4):1203

[53] Machado SDC et al. First report of Colletotrichum truncatum causing anthracnose in cassava in Brazil. Plant Disease. 2021;**102**(4):3745

[54] Khodadadi F et al. Identification and characterization of Colletotrichum species causing apple bitter rot in New York and description of C. noveboracense sp. nov. Scientific Reports. 2020;**10**(1):1-19

[55] Eaton MJ et al. Diversity and cross-infection potential of Colletotrichum causing fruit rots in mixed-fruit orchards in Kentucky. Plant Disease. 2021;**105**(4):1115-1128

[56] Zhang YL et al. The R2R3 MYB transcription factor MdMYB30 modulates plant resistance against pathogens by regulating cuticular wax biosynthesis. BMC Plant Biology. 2019;**19**:362. DOI: 10.1186/S12870-019-1918-4

[57] Oo MM, Yoon HY, Jang HA, Oh SK. Identification and characterization of Colletotrichum species associated with bitter rot disease of apple in South Korea. Plant Pathology Journal. 2018;**34**(6):480-489

Chapter 3

New and Emerging Disease Threats to Forest Plantations in Sarawak Borneo, Malaysia

Annya Ambrose, Jack Liam and Razak Terhem

Abstract

The planted forest area in Sarawak is the largest planted forest in Malaysia, which has been developed since 1997 to sustain the decline in the production of natural forests. As of December 2021, the total area of plantation forests reached 551,704 hectares (ha), dominated by fast-growing exotic species mainly Acacia species (55%), Falcataria moluccana (15%) and Eucalyptus (14%). The study showed Acacia was infected with red root rot disease of Ganoderma philippii and brown root rot of Phellinus noxius, Ceratocystis wilt disease caused by Ceratocystis. fimbriata sensu stricto (s.s) complex and pink disease caused by Erythricium salmonicolor, while F. moluccana was infected by gall disease namely Uromycladium falcatarium. Eucalyptus pellita diseases were infected namely by G. philippii red root rot disease, stem canker disease caused by Botryosphaeriaceae pathogen and bacterial wild disease caused by Ralstonia solanacearum. Ceratocystis wilt disease of Acacia mangium shows disease incidence (DI) accounted at 68% (serious) as compared with other diseases observed in this study. This will be the first baseline study that is conducted to observe and assess the diversity of the present, new and emerging pathogens and the damage they cause to exotic planted species of Sarawak.

Keywords: Acacia diseases, baseline study, disease incidence, Eucalyptus diseases, fast-growing exotic species, *Falcatria* disease

1. Introduction

The timber and timber-based products are the fourth largest contributor to Sarawak's export earnings after natural gas, petroleum and palm oil products. The timber industry, in particular, contributed to RM6.31 billion (USD 1.43 billion) in royalty, premium and cess revenue over a ten-year period, from 2010 to 2021. In addition, it also provides employment, creates business opportunities and facilitates road access for the rural communities that subsequently support eradication of poverty and sustainable development of the country.

Log supply from the natural forests peaked in the 1990s at about 20 million m^3 and declined to 4.07 million m^3 in 2019. The concern for the sustainability of natural forests has become a major topic of discussion these days. Sarawak is committed to conserving a healthy forest environment for the long term, including biodiversity conservation and ecosystem functioning with an aim to strike balance between the development and economic needs of the nation. Sarawak has a total land area of approximately 12.4 million hectares of which 7.7 million hectares or 62 per cent is under forest cover. Under the 'Sarawak Land Use Policy,' 7 million hectares have been allocated for sustainable forestry and conservation. Out of these, 6 million hectares are Permanent Forest Estates (PFEs) and 1 million hectares are Totally Protected Areas (TPAs). The remaining 5.4 million hectares are for agriculture and other development purposes. Under the Sarawak Forest Policy, the key component is sustainable forest management (SFM). SFM is implemented based on the guidelines in the Forest Statement Policy 1954.

Sarawak embarked on large-scale forest plantations in 1996 as a long-term strategy to provide a new source of wood material for the State's wood-based industries. Hitherto, Sarawak is the main player in Malaysia's forest plantation industry in the county where the total area of its plantation forests reached 551,704 hectares with three (3) main species planted, dominated by *Acacia* (55%), *Falcataria moluccana* (19%) and *Eucalyptus* (14%). Sarawak hopes to establish one (1) million hectares of industrial forest by 2025 by planting fast-growing species. It is hoped to produce 15 to 25 million m^3 a year when the target is achieved. In addition, the planted forest will relieve pressure on logging in natural forests in terms of both area size and intensity.

Currently, the forest plantation area could only produce an average production yield per ha of 70–120 m^3 assuming fully stocking at the age of 8 years trees. However, to better secure the future wood supply that Sarawak needs for its domestic and international markets, the production yield per ha should be increased to an average of 120–210 m^3 to be on par with other timber-producing countries such as Vietnam, China and Indonesia. One of the major attributes that had been identified in contributing to the poor growth and performance is the occurrences of disease attacks as well as the unavailability of genetically improved planting materials that are disease-resistant.

The research effort on forest health in Sarawak is still in the infant stage; however, its capacity has increased over the past decade. Surveys and basic quantitative evaluations were conducted during the period 2015–2020 in response to the plantation development observed in Sabah and Indonesia. Surveys and evaluations for forest health are crucial, and they should be conducted immediately and systematically in order to obtain related information such as disease incidence, signs and symptoms, contributing factors, and pathogens as well as silviculture treatments.

This information is expected to be used by stakeholders and policy-makers in developing an appropriate disease management strategy to control the disease infection on their plantations. Hence, this is the first baseline study conducted to identify and evaluate the incidence and the damages of the known major diseases as well as the emerging disease that could pose a threat to the vitality of forest plantation tree species in Sarawak.

The significant information gathered from this study will definitely provide an overview of the risk and threats of the pathogens and the management options that could be undertaken for a countermeasure.

2. Materials and methods

2.1 Forest health surveillance (FSH)

Forest health surveillance (FHS), a formal and regular inspection of the planted forest, was performed in this study through the establishment of permanent sampling plot (PSPs) surveillance and monitoring plots to assess the disease incidences and occurrences. FSH was undertaken between the period years 2015–2020, in various plantations selected randomly from the four forestry administration regions of Sarawak namely Miri, Bintulu, Sibu and Kuching (**Figure 1**). Field observations were carried out on forests planted with three major exotic tree species (**Table 1**).

We relied on information gathered by local foresters as well as our own observations to select the surveyed plantations in each studied forest region. Basically, the largest operational forest plantations were selected to better represent each forest region.

The plots were established by 30 × 30 m plot (the size of a plot varies in order to obtain a minimum of 100 living trees per plot) * or alternatively by line transect with 10 planting rows and 10 trees per rows approach, which subsequently 100 trees will be assessed, assuming every tree were still alive (full stocking). All the trees assessed were labelled accordingly. GPS coordinates were recorded at every corner of the plot

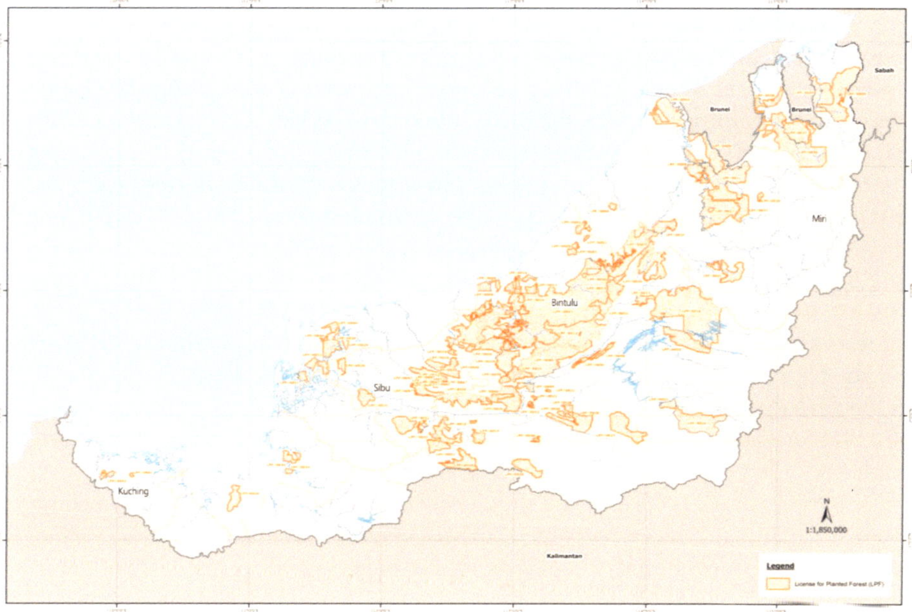

Figure 1.
Map of Sarawak showing the distribution of forest tree plantation which are highlighted areas with red border.

Host	Area in 2021 (ha)	Planting region	Native/ Exotic	Potential end-uses
Acacias	301,653.30	Kuching, Miri, Bintulu, Sibu	Exotic	Veneer, sawn timber, wood chips
Eucalyptus pellita	78,834.51	Kuching, Miri, Bintulu, Sibu	Exotic	Solid timber, sawn timber, timber for furniture
Falcataria molluccana	106,276.20	Kuching, Miri, Bintulu, Sibu	Exotic	Wood chips, veneer

Table 1.
Details of forest plantation species that were surveyed for disease pathogens.

established and marked clearly with a peg. The plots were established at least 20 m from the boundary of plantation, roads or forest gaps. Details of plots are shown in **Table 2**.

2.2 Disease assessment, sampling and disease incidence (DI)

Trees with signs and symptoms such as the presence of any fungal fruiting bodies, swelling and cracking of the stems, branches and canopy appearances (i.e. wilting, yellowing, or defoliations) were observed and thoroughly recorded. Various observations were also taken, including the presence of animal damage, and appearances of insect/pests (i.e. termites' mounds, defoliators, borers or boreholes).

Samples of diseased leaves, branches, stems and roots were taken from diseased trees and collected for laboratory examination. The surface of cankers on stems was examined on the field using hand lenses for dark pycnidia [1] or any signs of fruiting bodies. Trees showing symptoms of branch death, wilting foliage, bark discolouration or sap exudation typical of Ceratocystis as well as showing signed of wood boring insect activities and sweet gaseous smell of pineapple were examined by chopping into the woody xylem at the base of the tree (most infections occur through the root system) or further up the stems for the characteristic xylem discolouration. Longitudinal strips of wood xylem (approximately 0.5 cm thick and around 10 to 15 cm long) were then cut from the discoloured xylem. In some cases, diseased trees were also felled and were cut into several sections measuring between 10 and 20 cm in length. The samples were then placed into paper bags before being transported to the Forest Pathology Laboratory, Industrial Forest Research Centre (IFRC), Kuching. The samples were stored at 4°C prior to fungal isolation.

Below ground variables were recorded for each tree on roots exposed to 20–50 cm from the base of the tree that exhibits symptoms of root rots. Variables recorded included are the presence or absence of any kind of root rot (red, brown, white) as well as the presence or absence of insect pests such as termites' tunnelling activities. Samples of insect pests were also collected for further identification.

The surveys aim to estimate the number of trees with any signs and symptoms of disease infection and thus determine the disease incidences (DI).

The DI were calculated by the following formula:

$$\mathrm{DI} = \left(\frac{n}{N}\right) x \; 100\%. \tag{1}$$

Where:

n = number of infected standing tree

N = total number of standing trees in one plot.

The disease incidences were then evaluated based on the indicator and rating as guided in **Table 3**. Values DI were reported in the results as ranges and are based on the plot data over the years of plot assessment.

2.3 Fungal isolation and identification

2.3.1 Isolation of Ceratocystis wilt disease

The carrot baiting method with minor modification was carried out to isolate Ceratocystis [3]. Pieces of the diseased stem were placed between two slices of fresh

No	Plot ID	Species	Region*	Age (years)	Rotation	Spacing (m)	Types of stand**
1	SGAM1	*A. mangium*	BTU	6	1	3 × 3	CS
2	SGAM 2	*A. mangium*	BTU	6	1	3 × 3	CS
3	SGAM 3	*A. mangium*	BTU	1.5	2	3 × 3	CS
4	SGAM 7	*A. mangiun*	BTU	10	1	3 × 3	CS
5	SGAM 9	*A. mangium*	BTU	1	2	3 × 3	CS
6	PDAM01	*A. mangium*	BTU	3	2	3 × 3	CS
7	PDAM02	*A. mangium*	BTU	3	2	3 × 3	CS
8	GPP-SFC01	*A. mangium*	BTU	<1	2	3 × 3	CS
9	GPP-SFC02	*A. mangium*	BTU	<1	2	3 × 3	CS
10	SFC-DK-001	*A. mangium*	BTU	2	2	3 × 3	CS
11	SFC-DK-002	*A. mangium*	BTU	2	2	3 × 3	CS
12	SFC-DK-003	*A. mangium*	BTU	2	2	3 × 3	CS
13	SFC-DK-004	*A. mangium*	BTU	2	2	3 × 3	CS
14	SFC-DK-005	*A. mangium*	BTU	2	2	3 × 3	CS
15	SFC-DK-006	*A. mangium*	BTU	2	2	3 × 3	CS
16	SFC-DK-007	*A. mangium*	BTU	<1	2	3 × 3	CS
17	SFC-DK-008	*A. mangium*	BTU	<1	2	3 × 3	CS
18	SFC-DK-009	*A. mangium*	BTU	<1	2	3 × 3	CS
19	SFC-DK-010	*A. mangium*	BTU	<1	2	3 × 3	CS
20	BT 1–2016	*A. mangium*	BTU	3	2	3 × 3	CS
21	BT2–2016	*A. mangium*	BTU	2	2	3 × 3	CS
22	SBW 1–2016	*A. mangium*	SBW	2	2	3 × 3	CS
23	BT1–2017	*A. mangium*	BTU	2	2	3 × 3	CS
24	Lws 2-RND	*A. mangium*	MY	1	1	3 × 3	CS
25	Lws-3-RND	*A. mangium*	MY	1	1	4 × 3	CS
26	Lws 4-RND	*A. mangium*	MY	1	1	5 × 3	CS
27	Plot 1AH	*A. hybrid*	SBW	2	2	3 × 3	CS
28	Plot 2AH	*A. hybrid*	SBW	2	2	3 × 3	CS
29	SBW 1-RND	*A. mangium*	SBW	2	1	3 × 3	R&D
30	SBW 2-RND	*A. mangium*	SBW	2	1	3 × 3	R&D
31	SBW 3-RND	*A. mangium*	SBW	2	1	3 × 3	R&D
32	AM4	*A. mangium*	KCH	12	1	3 × 3	CS
33	EP1	*E. pellita*	KCH	11	1	3 × 3	CS
34	EP2	*E. pellita*	KCH	11	1	3 × 3	CS
35	EP3	*E. pellita*	KCH	4	1	3 × 3	CS
36	EP6*	*E. pellita*	KCH	4	1	3 × 3	CS
37	EP5	*E. pellita*	KCH	13	1	3 × 3	CS
38	4	*E. pellita*	BTU	3	2	3 × 3	CS

No	Plot ID	Species	Region*	Age (years)	Rotation	Spacing (m)	Types of stand**
39	5	*E. pellita*	BTU	3	2	3 × 3	CS
40	8	*E. pellita*	BTU	8	2	3 × 3	CS
41	10	*E. pellita*	BTU	4	1	3 × 3	CS
42	11	*E. pellita*	BTU	2	2	3 × 3	CS
43	PDEP01	*E. pellita*	BTU	2	2	3 × 3	CS
44	Plot 5 Bt	*E. molluccana*	SBW	2	2	3 × 3	CS
45	Plot 5 Bt	*F. molluccana*	SBW	2	2	3 × 3	CS
46	SFC-WTK003	*F. molluccana*	SBW	<1	2	3 × 3	CS
47	PD-AF-01	*F. molluccana*	BTU	3	2	3 × 3	CS
48	PD-AF-02	*F. molluccana*	BTU	3	2	3 × 3	CS
49	GPP-SFC03	*F. molluccana*	BTU	<1	2	3 × 3	CS
50	GPP-SFC04	*F. molluccana*	BTU	<1	2	3 × 3	CS
51	SFC -Bt-0018	*F. molluccana*	BTU	4	2	3 × 4	R&D
52	SFC-SY-001	*F. molluccana*	MY	2	2	3 × 3	CS
53	SFC-SY-002	*F. molluccana*	MY	2	2	3 × 3	CS
54	SFC-SY-003	*F. molluccana*	MY	2	2	3 × 3	CS
55	SFC-SY-004	*F. molluccana*	MY	2	2	3 × 3	CS
56	SFC-LimbaJ	*F. molluccana*	MY	2	2	3 × 3	R&D

Forestry Region (Figure 1**): BTU, Bintulu; SBW, Sibu; KCH, Kuching; MY, Miri; R&D.*
***Types of stand: R&D, Research and Development; CS, Commercial Stands.*

Table 2.
Details of permanent sampling plots established and evaluated.

Indicator	No	Light	Medium	Serious
Diseases incidence	Affect less than 10% of standing trees	Affect 10–29% of standing trees	Affect 30–59% of standing trees	Affect more than 60% standing trees

Table 3.
Indicator and ratings for evaluating the impact of forest health status in Sarawak [2].

carrot and incubated in plastic containers that served as moisture chambers for 4–6 days. After 5–8 days, a pale orange ascospore mass appeared on the tip of the perithecia was then picked with a sterile syringe needle and transferred into a Petri dish plated with Potato Dextrose Agar medium and incubated at 26°C. All the isolates were identified based on the analyses of morphological characteristics [3–5]. The characteristics and structures of the culture, such as perithecium, ostiolar hyphae, ascospore and conidia (barrel-shaped and cylindrical-shaped) (μm) of all the fungal isolates, were measured and described. Morphological identification for this study was conducted by using a Leica compound microscope ICC50HD with a laptop workstation for data & image acquisition and a stereo microscope of Olympus SZ61 (SZ61TR-2X-DI-TP-118000A-SET) at Forest Pathology Laboratory, IFRC, Kuching.

2.3.2 Isolation of root rot disease

Direct isolation was carried out to isolate the pathogen of root rot disease. Root tissues bearing disease symptoms were cut into several pieces approximately 3 mm × 3 mm in size and were placed on malt extract agar (MEA) and incubated at room temperature for the fungus to grow. The cultures were then transferred to MEA for morphological studies.

2.3.3 Isolation canker disease

Segments of symptomatic plant parts were incubated in moist chambers for 2–3 days to induce the development of fruiting structures. These were then transferred to MEA and incubated at 25°C. Isolation from symptomatic tissue was also made directly onto MEA. Isolations were made onto MEA from fruiting structures occurring on cankered stems [6, 7].

3. Results

3.1 Disease incidence (DI)

In this study, a total of 56 PSPs comprising 32 of *Acacia*, 17 of *Eucalyptus pellita* and 14 of *F. mollucana* have been established and assessed with approximately 5600 trees of the three major species planted have been evaluated, data recorded and analysed. The study showed that DI was varied between tree species.

3.1.1 Acacia

Approximately 3200 *Acacia* stand aged between less than 1-year-old and 12-year-old were assessed in this study with three prominent diseases observed; Ceratocystis wilt disease, root rot disease and pink disease. Based on **Table 4**, Ceratocystis wilt disease was the most devasted and important disease observed. DI was recorded the highest with mortality recorded at 68% in the Bintulu region based on disease indicators rating as depicted in **Table 2**, and the DI rating is categorised at a serious level. The disease was observed in stands less than 1-year-old also in the Bintulu region with DI recorded at 1.9%. The DI is predicted to increase over time with rotation length. The disease was not observed in Miri and Sibu region; however, DI of 3% of the disease was documented in Kuching region in 12 -year-old *A. mangium*.

The disease incidence of root rot was recorded between 0 and 37% in trees aged less than 1 year old to 10 year old. 10 years of plantation of *Acacia mangium* in the Bintulu region recorded the highest incidence of the disease, and no incidences were recorded in the young stand. The disease rating based on the highest DI is light. Root rot disease seems to be the most encountered disease in most of the plots, with 17 out of 32 PSPs observed with the disease.

The pink disease was only observed in Sibu region and had affected 2-year-old taxa seed source trial plot established by Forest Department Sarawak. DI of pink disease was recorded at 5%.

No	Plot ID	Species	Region*	Disease Incidence (DI)			Age (years)	Rotation	Spacing (m)	Types of stand**
				Root rot	Ceratocystis wilt	Pink disease				
1	SGAM1	*A. mangium*	BTU	22	0	0	6	1	3×3	CS
2	SGAM 2	*A. mangium*	BTU	22	0	0	6	1	3×3	CS
3	SGAM 3	*A. mangium*	BTU	29	0	0	1.5	2	3×3	CS
4	SGAM 7	*A. mangiun*	BTU	37	0	0	10	1	3×3	CS
5	SGAM 9	*A. mangium*	BTU	6	0	0	1	2	3×3	CS
6	PDAM01	*A. mangium*	BTU	17	13	0	3	2	3×3	CS
7	PDAM02	*A. mangium*	BTU	0	68	0	3	2	3×3	CS
8	GPP-SFC01	*A. mangium*	BTU	0	0	0	<1	2	3×3	CS
9	GPP-SFC02	*A. mangium*	BTU	0	0	0	<1	2	3×3	CS
10	SFC-DK-001	*A. mangium*	BTU	9.5	0	0	2	2	3×3	CS
11	SFC-DK-002	*A. mangium*	BTU	3.6	7.1	0	2	2	3×3	CS
12	SFC-DK-003	*A. mangium*	BTU	0	2	0	2	2	3×3	CS
13	SFC-DK-004	*A. mangium*	BTU	0	0	0	2	2	3×3	CS
14	SFC-DK-005	*A. mangium*	BTU	3.8	10	0	2	2	3×3	CS
15	SFC-DK-006	*A. mangium*	BTU	1.4	0	0	2	2	3×3	CS
16	SFC-DK-007	*A. mangium*	BTU	0	4	0	<1	2	3×3	CS
17	SFC-DK-008	*A. mangium*	BTU	0	15.5	0	<1	2	3×3	CS
18	SFC-DK-009	*A. mangium*	BTU	0	12.2	0	<1	2	3×3	CS
19	SFC-DK-010	*A. mangium*	BTU	0	1.9	0	<1	2	3×3	CS
20	BT 1–2016	*A. mangium*	BTU	8.3	35.2	0	3	2	3×3	CS
21	BT2–2016	*A. mangium*	BTU	3	3.5	0	2	2	3×3	CS
22	SBW 1–2016	*A. mangium*	SBW	1.2	0	0	2	2	3×3	CS

DOI: http://dx.doi.org/10.5772/intechopen.107027

No	Plot ID	Species	Region*	Disease Incidence (DI)			Age (years)	Rotation	Spacing (m)	Types of stand**
				Root rot	Ceratocystis wilt	Pink disease				
23	BT1–2017	*A. mangium*	BTU	7.2	41.2	0	2	2	3 × 3	CS
24	Lws 2-RND	*A. mangium*	MY	0	0	0	1	1	3 × 3	CS
25	Lws-3-RND	*A. mangium*	MY	0	0	0	1	1	4 × 3	CS
26	Lws 4-RND	*A. mangium*	MY	0	0	0	1	1	5 × 3	CS
27	Plot 1AH	*A. hybrid*	SBW	1.2	0	0	2	2	3 × 3	CS
28	Plot 2AH	*A. hybrid*	SBW	1.2	0	0	2	2	3 × 3	CS
29	SBW 1-RND	*A. mangium*	SBW	0	0	5	2	1	3 × 3	R&D
30	SBW 2-RND	*A. mangium*	SBW	0	0	3	2	1	3 × 3	R&D
31	SBW 3-RND	*A. mangium*	SBW	0	0	2	2	1	3 × 3	R&D
32	AM4	*A. mangium*	KCH	3	3	0	12	1	3 × 3	CS

Forestry Region (see **Figure 1): BTU, Bintulu; SBW, Sibu; KCH, Kuching; MY, Miri; R&D.*
***Types of stand: R&D, Research and Development; CS, Commercial Stands.*

Table 4.
Details on DI (DI) of observed diseases recorded in Acacia.

No.	Plot ID	Species	Region*	Disease Incidence (DI)			Age (years)	Rotation	Spacing (m)	Types of stand**
				Root rot	Canker	Bacterial Wilt				
1	EP1	*E. pellita*	KCH		4		11	1	3 × 3	CS
2	EP2	*E. pellita*	KCH		3		11	1	3 × 3	CS
3	EP3	*E. pellita*	KCH		2		4	1	3 × 3	CS
4	EP6*	*E. pellita*	KCH				4	1	3 × 3	CS
5	EP5	*E. pellita*	KCH		2		13	1	3 × 3	CS
6	4	*E. pellita*	BTU				3	2	3 × 3	CS
7	5	*E. pellita*	BTU				3	2	3 × 3	CS
8	8	*E. pellita*	BTU	6			8	2	3 × 3	CS
9	10	*E. pellita*	BTU	6			4	1	3 × 3	CS
10	11	*E. pellita*	BTU		7		2	2	3 × 3	CS
11	PDEP01	*E. pellita*	BTU				2	2	3 × 3	CS
12	PDEP02	*E. pellita*	BTU				2	2	3 × 3	CS
13	Plot 3 EP	*E. pellita*	SBW		1.2		2	2	3 × 3	CS
14	Plot 4 EP	*E. pellita*	SBW				2	2	3 × 3	CS
15	SFC-WTK1	*E. pellita*	SBW				<1	2	3 × 3	CS
16	SFC-WTK2	*E. pellita*	SBW				<1	2	3 × 3	CS
17	SFC-BT	*E. pellita*	BTU			3	2	2	3 × 3	R&D

Forestry Region (see* ***Figure 1**): BTU, Bintulu; SBW, Sibu; KCH, Kuching; MY, Miri; R&D.*
***Types of stand: R&D, Research and Development; CS, Commercial Stands.*

Table 5.
Details on disease incidence (DI) of observed diseases recorded in E.pellita.

3.1.2 E. pellita

Approximately 1700 *E. pellita* stands aged between less than 1-year-old and 13-year-old were assessed in this study with the establishment of 17 PSPs as detailed in **Table 5**. Canker disease was found in 6 PSPs with the highest DI of 7% recorded in 2-year-old stand in Bintulu region. Out of the 6 PSPs, mostly canker incidences were found in Kuching with 11% DI accumulatively. Root rot was observed in the Bintulu region with accumulative DI of 12% in 2 PSPs located within the same area. Bacterial wilt disease with a DI of 3% was encountered in one PSP established within R&D tree breeding plot at Bintulu. No other PSPs recorded such disease.

3.1.3 F. molluccana

A total of 14 PSPs with comprises of approximately 1400 of *F. molluccana* stands of less than 1-year-old and 2- and 3-year-old were established and assessed for DI as

No.	Plot ID	Species	Region*	Disease Incidence (DI) %		Age (years)	Rotation	Spacing (m)	Types of stand*
				Gall Rust	Canker				
1	Plot 5 Bt	*F. mollucana*	SBW			2	2	3 × 3	CS
2	Plot 5 Bt	*F .mollucana*	SBW			2	2	3 × 3	CS
3	SFC-WTK003	*F. mollucana*	SBW			<1	2	3 × 3	CS
4	PD-AF-01	*F. mollucana*	BTU			3	2	3 × 3	CS
6	PD-AF-02	*F. mollucana*	BTU			3	2	3 × 3	CS
7	GPP-SFC03	*F. mollucana*	BTU		1	<1	2	3 × 3	CS
8	GPP-SFC04	*F. mollucana*	BTU			<1	2	3 × 3	CS
9	SFC -Bt-0018	*F. mollucana*	BTU	69.7		4	2	3 × 4	R&D
10	SFC-SY-001	*F. mollucana*	MY			2	2	3 × 3	CS
11	SFC-SY-002	*F .mollucana*	MY			2	2	3 × 3	CS
12	SFC-SY-003	*F. mollucana*	MY			2	2	3 × 3	CS
13	SFC-SY-004	*F. mollucana*	MY			2	2	3 × 3	CS
14	SFC-LimbaJ	*F. mollucana*	MY	1.9		2	2	3 × 3	R&D

Forestry Region (see **Figure 1): BTU, Bintulu; SBW, Sibu; KCH, Kuching; MY, Miri; R&D. Types of stand: R&D, Research and Development; CS, Commercial Stands.*

Table 6.
Details on disease incidence (DI) of observed diseases recorded in F. molluccana.

detailed in **Table 6**. Based on the survey, gall rust disease was the most important prominent disease observed with DI recorded at an alarming 69.7% in an R&D (taxa seed source trials) established at Bintulu. Miri region also recorded gall rust DI of 1.9% in stands less than 1-year-old. Another disease observed affecting *F. molluccana* was canker disease and attacked stands less than 1-year-old also in the Bintulu region with DI recorded at 1.9%. No diseases were observed in the Sibu region.

3.2 Symptoms and signs

3.2.1 Diseases of Acacia

3.2.1.1 Ceratocystis wilt disease of Acacia

Ceratocystis infections observed typically start on the lower stem and move up, causing dark staining of the woody xylem. However, we did observe some symptomatic trees with initial infection starting at the upper stem with dead foliage hanging throughout the tree or on some major branches. The infected blackish gummosis stem upon debarking showed dark brown to black discolouration in the woody xylem, radial pattern in pattern, which is a unique symptom of Ceratocystis wilt disease. Within every plot observed with Ceratocystis, all trees showed these wilt symptoms were noted for Ceratocystis wilt (**Figure 2a–d**). But it was further confirmed upon root excavation to differentiate root rot disease as the latter also exhibits symptoms such as wilting and severe defoliation.

Figure 2.
Ceratocystis wilt disease implicates damages to Acacia mangium *plantation in Sarawak (a–i); (a) blackish lower stem with foamy exudate at the upper stem and wilting and dying 2-year-old tree; (b) blackish streak-like inner bark due to Ceratocystis infection; (c) foamy exudates and slime typical of Ceratocystis infection on stem of the tree; (d) initial infection starting at the upper stem with dead foliage hanging throughout the tree; (e) dead tree with severe bark damage caused by squirrels with wood discolouration; (f) severe bark damage but tree observed free from Ceratocystis infection (g) traces of wood-boring insect holes on the stem;(h) Ambrosia beetle found under bark and the wood appeared with blackish lesion; (i) side view image of wood-boring insect identified as* Immanus desectus *(eggers) collected from the infested tree.*

Some symptomatic and dead trees as young as 6-month-old that were adjacent to the conservation or buffer area were observed with wound damage on the stem of juvenile trees inflicted by squirrels grazing (**Figure 2e**). These wounds would act as the point of entry for Ceratocystis pathogen to infect the trees. However, we did observe that some trees inflicted with these wounds but did not show any Ceratocystis symptoms (**Figure 2f**).

Wood-boring insects were also found present in infected trees, with wood infested exhibited blackish lesions typical of *Ceratocystis* pathogen (**Figure 2g–i**). The wood-boring insects collected from the tree were identified as those of ambrosia beetle species.

Culture characteristics and morphology of Ceratocystis isolates from *A. mangium* were observed and analysed based on morphological characteristics, such as colony form, mycelium colour and reverse media colour [3–5], were further identified as *C. fimbriata sensu stricto (s.s) complex*, having the characteristic of olive-green colonies with the underside of the cultures was light grey at the margin and became darker towards the centre and typical pineapple-fruit odour. The isolates of *C. fimbriata* showed superficial or submerged in the substrate perithecia, with colours ranging from brown to black both at the base and neck of the perithecium (**Figure 3c**). They had globose to sub-globose ascomata with long necks and typical divergent ostiolar hyphae at their tips (**Figure 2**). Teleomorph and anamorph structures were produced within 2 weeks on MEA cultures (**Figure 3d**).

3.2.1.2 Root rot disease of Acacia

The trees showed symptoms typical of red root rot including dying trees at the periphery of rot centres that had wilting leaves. Most of the affected trees were dead stands that exhibited chlorotic/yellowing foliage and loss of foliage.

The diseased trees were observed to be clustered in patches that were roughly circular. The trees with symptoms of root rots were adjacent to one another. Fruit bodies were only observed on the stem of some of the trees. Types of root disease were distinguished by the colour of infected roots, red root rot, brown root rot and white root rot.

Underground symptoms observed include the presence of red-coloured rhizomorphs with blackish exudates (**Figure 4c**) on the surface of the roots and white mycelium (denote as *My* in **Figure 4d**). Yellow mottles of the foliage with patches in the majority of the infected *A. mangium* trees were easy to recognise as typical of red root rot and further identified as *G. philippii*. Trees discovered to be infected with brown root rot were observed with soft, fibrous and totally rotten roots and sandy particles adhering to the rhizomorph (**Figure 4e**). The causal pathogens identified were *G. philippii* and *Phellinus noxius*.

3.2.1.3 Pink disease of Acacia

Pink disease in *A. mangium* was observed both in the base of the main stem and branches and twigs. Trees observed with pink disease (**Figure 5**) initially exhibited silky-like whitish mycelium on the surface of the bark, known as the 'cobweb' stage, infecting the branch and stem. Pinkish to salmon cobweb-like structures were observed soon after known as the pustule stage, and subsidiaries symptoms of wilting of the branch's foliage and subsequently turned brown and branches that extending to all parts of the plant until dieback observed by death in the canopy. Infected branches and stems showed sign of bark crack and scaled bark. Where infection was observed centralised in certain areas of the stem or bark, cankers and wounds developed. In this

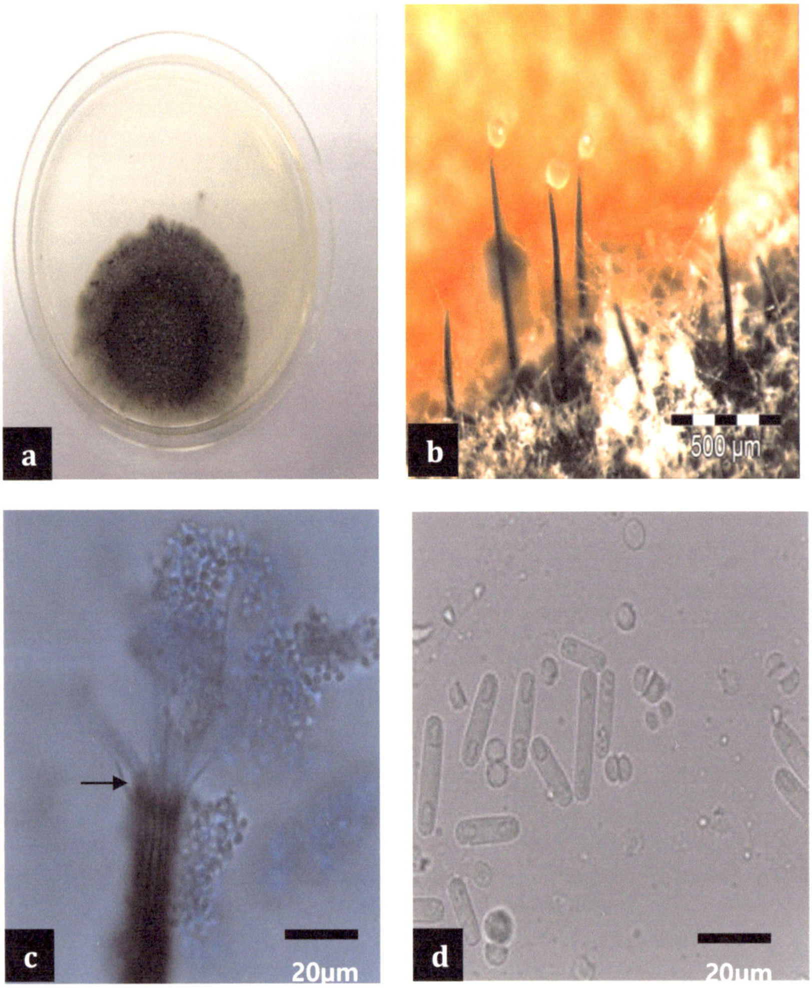

Figure 3.
Microscopic structures of identified Ceratocystis fimbriata *obtained from* Acacia mangium *plantations in Sarawak* ***(a–d)****; (a) culture of* C. fimbriata *grown on MEA appeared cotton-like, sparse, initially white, later turning to olive grey with white to grey at the margin after 5–7 days with banana-like aroma and salmon-like shiny globose spores; (b) Salmon-like colour ascospores mass extruded at the tips of the long-necked black fruiting bodies (perithecia); (c) long, divergent ostiolar hyphae (arrow); (d) spores of* C. fimbriata., Thelaviopsis *sp.-elongated, cylindrical condia of* Thelaviopsis *sp. (imperfect state) and hat-shaped ascoscopres of* C. fimbriata *(teleomorph or perfect state).*

survey, we observed severe infection in one PSP in Sibu region and inflicted mortality of the tree stands. The symptoms described above showed similarity with the symptoms described [8] and thus, the pathogen of pink disease was identified as *Erythricium salmonicolor*.

3.2.2 Diseases of E. pellita

3.2.2.1 Root rot disease of E. pellita

Root rot disease infecting *E. pellita* in this study was observed to be very detrimental and trees dying in patches. The first signs typical of root rot disease attacked

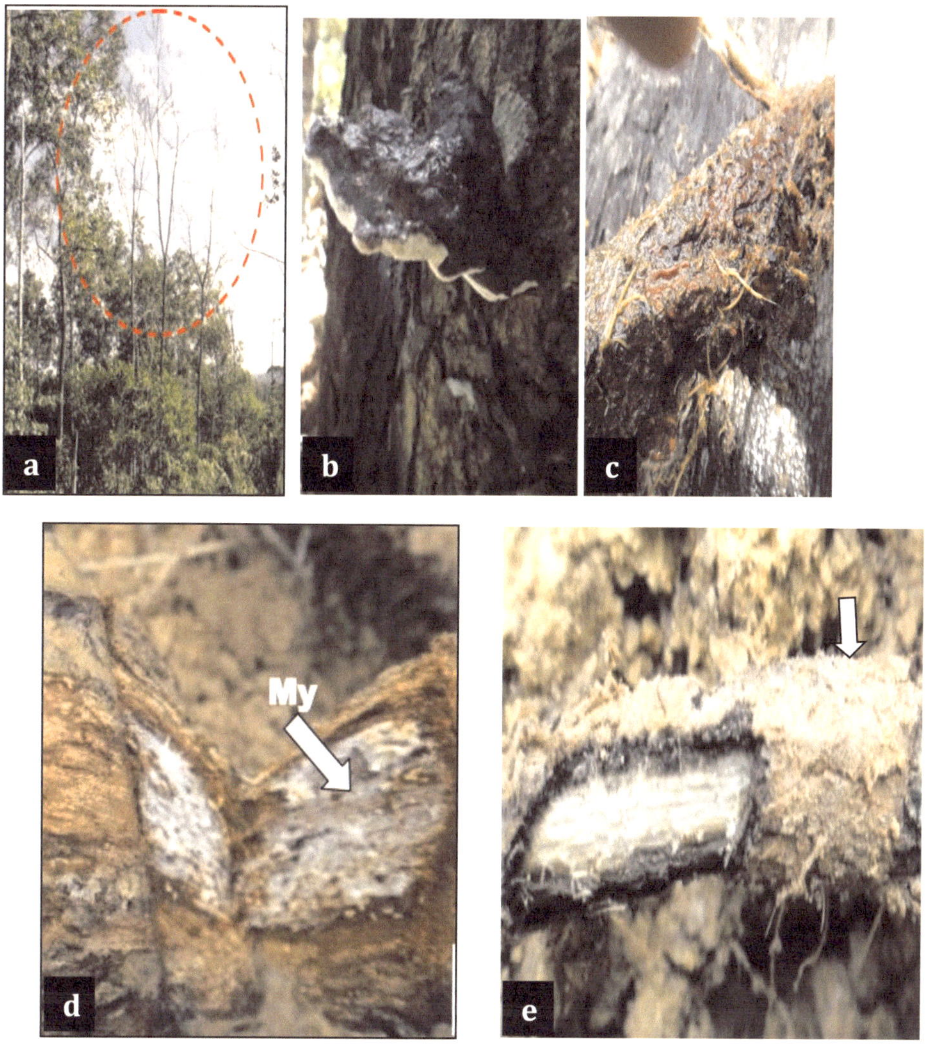

Figure 4.
Root rot disease of Acacia mangium*; (a) 10-year-old trees with severe defoliation and open gaps are the most advanced symptoms of the disease; (b) Ganoderma fruiting body on infected trees; (c) red-coloured leathery rhizomorphs with blackish exudates on infected roots; (d)whitish mycelium on the under bark; (e) Brown root rot caused by the pathogen* P. noxious *with soft roots, fibrous and totally rotten and sand like soil (arrow adhering to the rhizomorphs).*

were yellowing of the foliar, and falling with twigs and branches die. The whole crown seems to be burnt-like (**Figure 6a**) and subsequently, the tree died with the tree became bare. The base of the stem and roots appeared rotten, and discolouration of bark stem was observed with fine white threads typical mottled pattern of mycelia growth below the bark (**Figure 6b**). Fruit body body's shape is like a fan, semicircular, hard and woody and the bottom surface is white; the upper middle is dark brown and measured around 50 cm was seen growing on the stem's base. The roots were then uprooted revealing red leathery rhizomorph covering the root system. Following the symptoms observed especially the characteristic of the fruiting found on the tree, the pathogen was identified as *G. philippii* [9].

Figure 5.
Erythricium salmonicolor *pink disease in* a.mangium. *(a)* E. salmonicolor *'cobweb' stage of infection.; (b)* E. salmonicolor *'pustule stage; (c) incrustation pink cobweb-like colonies on upper stems; (d) patches of tree death caused by the pink disease pathogen* E. salmonicolor.

3.2.2.2 Canker disease of E. pellita

Symptoms of canker disease of *E. pellita* varied in each region where it was detected. Canker found in 13- to 11-year-olds (**Figure 7a–e**) in Kuching was found some at the base of the stem and others slightly higher up the stem (**Figure 7b**). The cankered stem characteristics are swollen and misshapen and the bark fissured with traces of drying blackish gummosis, but all trees were still alive without any sign of dying. The canker found in 2-year-old PSP trees stand in Bintulu showed characteristic of the whole bark cracking with stems blackened and leaves yellowing and wilting with mortality recorded. Based on the different symptoms observed in PSPs assessed, there were possibilities that more than 1 pathogen caused this disease. A small portion of the bark of one tree in 2-year-old stand in Bintulu region was collected for

Figure 6.
Root rot disease of Eucalyptus pellita*; (a) 6-month-old tree with burned like foliar and blackish stem; (b) mycelium on under the bark of the basal stem; (c) Ganoderma root rot fruiting body; (d) red-coloured leathery rhizomorphs with blackish exudates on* E. pellita *root; (e) rotting and fibrous root system.*

examination in the laboratory. We managed to identify the pathogen as Botryosphaeriaceae based on the conidia and the culture characteristic grown on MEA as guided (**Figure 8**) [10, 11].

Based on silviculture regimes were not recorded by the management of the plantation. It was observed weeding was not incorporated into the silviculture regime.

3.2.2.3 Ralstonia solanacearum*—bacterial wilt of Eucalyptus*

The symptoms started with foliar exhibiting chlorotic, yellowish to reddish and gradually spreading to the branches started to wilt and trees will succumb to mortality. The inner stem exhibits vascular discolouration of blackish to brownish and bacterial ooze from the wood when observed using a hand lens (**Figure 9**). The

Figure 7.
Canker disease symptoms of Eucalyptus pellita*; **(a–e)** stem cankers on 11-year-old* E. pellita *(a) double' canker at basal and middle stem on 11-year-old; (b) close-up showed cracked and fissured bark on 11-year-old; (c) wood under bark seems to be healthy on 11-year-old; (d) canker developed from injury because of self-pruning and on 11-year-old (e) upper stem canker with swollen bark on 11-year-old; (f) huge canker swelling on 2-year-old* E. pellita *at Bintulu region.*

infected trees were chopped and burned. The possible pathogen would be *R. solanacearum* Smith based on disease symptom described [12, 13].

3.2.3 Disease of Falcataria molluccana

3.2.3.1 Uromycladium falcatarium*—gall rust of* F. molluccana

The disease was observed to cause severe damages and mortality to young trees of *F. mollucana*, especially in the Miri region. The disease causes severe damage to all

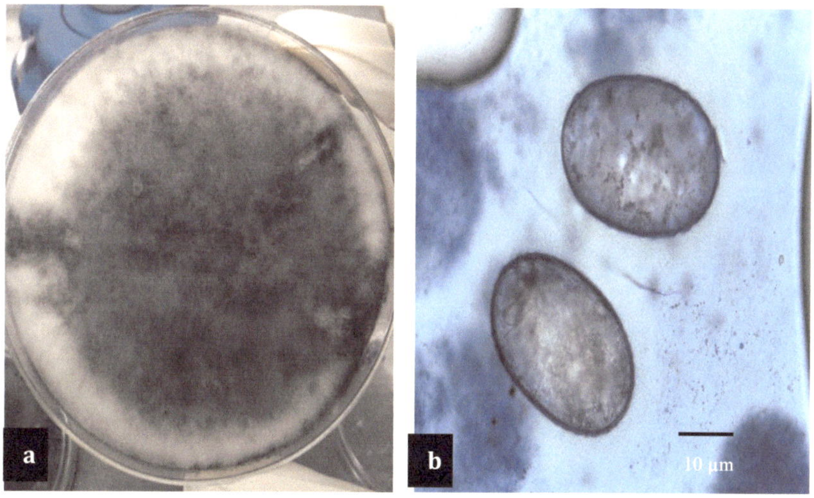

Figure 8.
Botryosphaeriaceae pathogen of canker disease of Eucalyptus pellita*. (a) Culture morphology on MEA; (b) conidia of* Botryosphaeria.

Figure 9.
Bacterial wilt sign and symptoms in Eucalyptus pellita*. (a) Wilting and drying branches symptom on 1-year-old eucalyptus; (b) colour changes turning blackish-brown on eucalyptus 1-year -old stem further symptoms (c) bacterial exudates (ooze) on wood.*

developmental stages of the plant, from seedlings in the nursery to mature trees in the field. Early symptom of infected trees observed was rigid, crooked bending stems, branches or shoot and greenish to reddish necrotic lesion (**Figure 10b**). The symptoms then developed to the development of large chocolate-brown, irregularly shaped, broccoli-like or whip-like galls on the stem and branches (**Figure 10c**).

Figure 10.
Gall rust of F. mollucana. *(a) Whole crown dieback and tree died in 2-year -old tree stands at Miri region; (b; c) the shape of gall in the twig; (d) spores with ridged longitudinal striations typical of* Uromycladium falcatarium *pathogen.*

The disease pathogen was identified as *U. falcatarium* based on the symptoms observed in this study as well as referred [14].

4. Discussion

This is the first baseline data of disease incidence (DI) of the current as well as emerging pathogen threats to plantation forests in Sarawak. It comprised about 5 years of field surveys as well as monitoring the health on a yearly basis. PSPs were established randomly across all the four regions of Sarawak and were representative of the 521702.27 ha of exotic tree species. Overall, seven species of plant pathogens were recognised as either current or emerging threats to the forestry sector. *Ceratocystis fimbriata* is of the greatest concern due to its significant increase in geographical range, the successive rotation and fear of spread or 'jumping host' from Acacia to other species such as *Eucalyptus*.

4.1 Diseases of Acacia

4.1.1 Root rot disease

This disease is considered one of the most important diseases of *A. mangium* plantations in Indonesia and Malaysia [15–17]. The pathogen builds up in every successive rotation and caused tree mortality in trees as young as 3-month-old in particular in the 2nd and 3rd rotation trees based on observation. Tree death can exceed 50% in some areas within less than 20 years of establishing the first rotation [18]. The trees with symptoms of root rots were adjacent to one another. It is a known fact that the root systems of trees are entangled with one another, and thus, one infected root will have higher possibility of infecting the roots of the tree next adjacent. As such, the trees infected with root rot appeared to be centralised and clustered in patches that were roughly circular. The nearer the planting distance, the probability spread of infection will increase exponentially.

Currently, the spread of root rot diseases is control by excavating and destroying the infecting trees as well as construction of drainage trenches filled then with fungicides, which will probably minimise the roots contact of infected trees to healthy trees adjacent. However, mostly all plantation owners commented that such practices is not economically viable and opted to just leaving the infected trees as it are. From general overview, root rot disease will always be a menace as the pathogen fungi, which are present in native forest left in infected stumps and roots when native forest is cleared for plantations, and these then act as inoculum sources and it will definitely increase with successive rotation [16]. Land preparation and silviculture management are the only way to minimise the inoculum build up as the wood debris left after planting will act as the food source of the fungi.

4.1.2 C. fimbriata complex-*Ceratocystis wilt disease*

The pathogen emerged as a force to be reckoned with, as with the concerns about the vitality and survival of Acacia plantations in Indonesia, Vietnam and Sabah since 2010. It had created havoc on the tree plantation industries in Indonesia and resulted in the replacement of approximately 600, 000 ha of *A. mangium* plantation with *E. pellita* and its hybrid [3, 19, 20]. The disease was first observed in Sarawak in early 2010 in the Bintulu region, infected not more than 10 trees adjacent to the roadsides and trees in the vicinity of the conservation area and buffer zone. Over the years, based on observation and feedback from plantation company personnel who worked on the ground, the number of infected trees is increasing and spreading to the region of Sibu, Miri and Kuching as well. Compared to root rot pathogen modes of spread through root contact, the spread of Ceratocystis pathogens through air and water could be considered rapid especially when the trees were inflicted with wounds made by pruning, animal damages such as squirrel observed in the Bintulu region. However, one most important discovery in some of the surveyed plots in Bintulu region was some infected trees showing sign of recovery from the infection with new sprouts growing out from the trees.

This could indicate resistance and/or tolerance of some of the individual trees. It was later found out that the plantation company has planted improved clones in their area. This justifies that breeding of disease should be actively pursued as a tool to help with the growing problem of invasive pests and pathogens that threaten our forests. However, resistance breeding is sometimes viewed as being too long term (5 years) and too expensive to be practical as lack of scientific understanding between planters and scientists in the matter concerned [21].

There are some assumptions that wood-boring insects could have the possibilities of harbouring the pathogen in their gut and acting as a vector of the disease; however, no concrete studies have been able to prove this theory. The attempted isolation of Ceratocystis from the carcass of the ambrosia beetle collected during the survey using the carrot bait method but none bore any traces of the pathogens, thus concluding that it did not harbour the fungus and thus not the vector in this specific study. Therefore, we presume the wood-boring beetle found is most probably due to secondary infestation caused by abiotic tree stress.

4.1.3 Pink disease

Currently, the pink disease caused by the pathogen *U. salmonicolor* was never been considered a major threat to Acacia plantations as compared to root rot and *Ceratocystis* wilt [22] disease as the reports of the pink disease were very few and it assumed that Acacia species particularly *A. crassicarpa* seems to be resistant to the fungal pathogen of *U. salmonicolor* [8, 23]. In Sarawak, the earliest occurrence of pink disease on *A. mangium* was recorded in 1979 and it reached epidemic proportion in 1987 [24]. The mortality inflicted by this pathogen observed in 2-year-old *A. mangium* indicated that this disease should be taken seriously. As the spores could travel by wind and water, it will be a threat to nurseries and early establishment of seedlings planted.

4.2 Disease of Eucalyptus

4.2.1 Canker disease

Several fungi are known to cause stem cankers in eucalyptus, among them *Botryosphaeria* spp., *Chrysoporthe deuterocubensis* and *Teratospaeria zuluense* (syn. *Coniothyrium zuluense*) [25]. *Chrysoporthe deuterocubensis* is a widespread and important pathogen of plantation eucalypts in the tropics. Basal cankers caused by *C. deuterocubensis* can extend several meters up the stem and have the ability to kill young trees. Where stems have been girdled, young trees may wilt and die suddenly during hot and dry weather [26–28].

Older trees of 11 to 12 years in the Bintulu region at the first rotation that exhibit symptoms of canker in this assessment are growing well and without any concern for mortality. The wood under bark seems to be healthy without any signs of lesion or discolouration although gummosis is present on the canker stem. This might be that trees normally could have survived the initial infection and only will develop basal swellings and severe bark cracking over brown necrotic sapwood [29]. However, younger trees of 2-year-old in this study recorded mortality. The trees are currently in 2nd rotation planting; thus, it could be concluded that DI and severity, as well as tree mortality in especially young *E. pellita* plantations, can thus only be expected to increase after the species has been grown for several rotations on the same site [16].

4.2.2 R. solanacearum*—bacterial wilt of Eucalyptus*

Bacterial wilt disease of Eucalyptus will pose a difficulty to contain as *R. solanacearum* sensu lato is known to be a destructive bacterial phytopathogen that is able to cause bacterial wilt in over 50 plant families growing in tropical, subtropical and some temperate areas globally [30]. Bacterial wilt arises from the blockage of the xylem tissue by bacterial growth, thereby resulting in wilt symptoms in the aerial parts of infected plants that ultimately die. The symptoms are very much similar to Ceratocystis wilt disease with the inner stem exhibiting blackish-brown. However, to distinguish between the latter and bacterial wilt disease, examination of the plant stem using a hand lens will be required to observe bacterial oozing (**Figure** 7). As with many pests and diseases, the technology of breeding and selection of disease and insect cultivars should be the key to minimising damage and ensuring the productivity of tree plantation industry [30, 31]. However, the process requires scientist to weigh many factors that include pathogen diversity, prevailing environmental conditions and the availability of material for breeding as well to make the management level understand that process is lengthy that requires at least 1 cycle rotation (8–9 years) planting and costly to begin with.

4.3 Disease of *F. mollucana*

Gall rust was found as the predominant disease of *F. mollucana* in part of Miri and Bintulu regions. It was first reported by some plantation owners back in 2012; however, the disease only affected the seedlings in the nursery. The source of infection was deduced from the seeds imported from the island of Jawa Indonesia. This could be related that Indonesia had experienced the major outbreak of the disease, which affected much of *F. molluccana* plantation estate in Java in 2010 [32].

The plantations affected were those of higher elevation of more than 152 m asl with spacing of 3 m^2 that could be considered to be a close gap between trees. The pathogens thrive in an environment with elevations ranging from 152 m asl to 975 m asl, in trees ages 1-year-old to 9 year-old, and spacing between 6 m^2 to 16 m^2 [33] and require fog or mist as well as high relative humidity to ensure infection [34] (m).

4.4 General overview

The loss of yield due implicated by insect pests and pathogens attacks could hamper the overall maximum production yield of the plantation forest. Sarawak, in recent years, has observed the dramatic losses of *A. mangium* plantation in its neighbouring country Sabah due to the incidence of a serious canker and wilt disease caused by *Ceratocystis acaciivora* [19], the conversion of 1 million hectares of the *A. mangium* plantation estate in Indonesia to *E. pellita* Muell. and related hybrids over the past 5 to 10 years as detailed by [20]. These critical developments have since propelled the forest plantation industry players in Sarawak to be more cautious and attentive to any signs of threats especially caused by fungal pathogens.

Insect pests have not been associated with widespread mortality or failure of any acacia plantations in Sarawak so far. However, an investigation survey conducted in early 2022 to determine the cause of mortality in one of *A. mangium* superbulk plantations in the Bintulu region revealed that termite's infestation further identified as *Coptotermes travians* appeared to be the most prevalent pest that had inflicted tree mortality in the area, which accounted for 24.1% in mortality of trees.

The tree plantations in Sarawak are currently in the second rotation planting but will soon enter the third rotation of planting, and the concerns of increasing pests and diseases attack will definitely be inevitable [35, 36]. There is growing concern that *A. mangium* may no longer be capable of producing commercial yields after three rotations [37]. Significant reductions in productivity have been reported with each successive rotation and reductions have frequently been associated with mortality caused by fungal pathogens [1].

Thus, Sarawak is looking for other alternative tree species to compliment or maybe even replace *A. mangium*. *E. pellita* and its hybrid are seen to be the next species of choice as Eucalyptus species are easy to hybridise as compared to Acacia species, which is due to the long time and the low productivity in multiplication due to the ageing effects [38]. By theory, hybrid clones combine fast growth, increased tolerance to pests and diseases, excellent rooting ability, as well as wood quality suitable for different uses [39].

Other tree species such as *F. mollucana* seem to be at a lower risk of pathogen attack; however, through observation and communication with planters, insect pests such as defoliators moth and borers will contribute to lower quality of the woods produced although mortality caused by these pests are expected of less concern. New exotic species of that Paulownia are slowly making a wave in Malaysia, and some planters in Sarawak are experimenting in planting this tree on a small scale.

The diagnosis and determination of the cause of any damage or disease on a tree can only be made when the trees are still alive. Once the tree is dead, it is very difficult if not impossible to determine the cause of the problem. Therefore, it is very important that tree health inspections and pest and disease surveys are carried out at regular intervals to enable recognition of early signs and/or symptoms of pests and disease infection [40, 41].

Once recognised, the initial stages and the development of symptoms can be followed until the death of the tree. This would go a long way towards better identifying the causes of tree mortality in the plantations. By so doing, instead of just encountering a dead tree during a random inspection, one would have detailed information about when the symptoms first appeared, their pattern of development and spread, mode and rate of spread, and thus a better idea of the agent(s) involved. It is, therefore, very encouraging that the management has decided to set up pest and disease monitoring transects in the young plantations.

As such, different strategies should be employed in better managing the threats of pests and disease, such as good silviculture, increasing the knowledge of proper identification of pathogens and pests threats among the planters, sharing of knowledge and more transparent in R&D data and information between different companies in Sarawak particularly, management consideration in acknowledging that R&D especially tree pest and disease surveillance, as basis it seems but the information gain is invaluable to predict the next outbreaks and to better contain the damages.

5. Conclusion

In view of the survey done in the period of 5 years through FHS, Ceratocystis wilt disease of *A. mangium* emerged as a major disease in hampering the progress of the

Sarawak forest plantation industry. Research effort must be intensified in making sure *A. mangium* could make it to the subsequent rotation. Although *E. pellita* and its hybrid are making a hype to replace *A. mangium* as the main species planted, diseases such as bacterial wilt and canker disease will be a major problem to planters later on if no countermeasure is readily in place.

Acknowledgements

This study was financed in part by the Sarawak State Government under the Ministry of Natural Resources and Urban Development (Rancangan Malaysia Ke-11) under the project titled 'Centre of Excellence for Planted Forest'. We acknowledge the support received from Datu Hamden Bin Haji Mohammad, Director of Forest Sarawak, and as well the assistance rendered by members of the Forest Health Unit (Erica Hadari, Lily Encharang, Jonnidi Suib and Jaies Chankul) throughout this study, and especially, the surveys and collection of diseased samples are highly appreciated.

Author details

Annya Ambrose[1,2], Jack Liam[2] and Razak Terhem[1*]

1 Laboratory of Forest Pathology and Tree Health, Department of Forestry Science, Faculty of Forestry and Environment, Universiti Putra Malaysia, Selangor, Malaysia

2 Laboratory of Forest Pathology, Industrial Forest Research Centre, Forest Department Sarawak, KM20 Jalan Puncak Borneo, Sarawak, Malaysia

*Address all correspondence to: razakterhem@upm.edu.my

References

[1] Oliveira LSS, Harrington TC, Ferreira MA, Damacena MB, Al-Sadi AM, Al-Mahmooli IHS, et al. Species or genotypes? Reassessment of four recently described species of the Ceratocystis wilt pathogen, *Ceratocystis fimbriata*, on *Mangifera indica*. Phytopathology. 2015;**105**:1229-1244. DOI: 10.1094/PHYTO-03-15-0065-R

[2] Sri R, Su SL, Nor AAS. *Uromycladium tepperianum*, the gall rust fungus from *Falcataria moluccana* in Malaysia and Indonesia. Mycoscience. 2010;**51**: 149-153. DOI: 10.1007/s10267-009-0022-2

[3] Syazwan SA, Mohd-Farid A, Wan-Muhd-Azrul WA, Syahmi HM, Zaki AM, Ong SP, et al. Survey, identification, and pathogenicity of *Ceratocystis fimbriata* Complex associated with wilt disease on *Acacia mangium* in Malaysia. Forests. 2021;**12-18**:1782. DOI: 10.3390/f12121782

[4] Nícolas MDB, Henrique DSSD, Caroline DBB, Celso GA, Álvaro FDS. Morphophysiological characterization of Ceratocystis fimbriata isolates from yerba mate. Ciência Rural. 2021;**51**(3): 1-5. DOI: 10.1590/0103-8478cr20200579

[5] Suwandi S, Chandra I, Harman H, Abu U, Khoirotun DA. Identification and characterization of Ccratocystis fimbriata causing lethal wilt on the Lansium tree in Indonesia. Plant Pathology Journal. 2021;**37**(2):124-136. DOI: 10.5423/PPJ.OA.08.2020.0147

[6] Alemu G, Roux J, Slippers B, Wingfield MJ. Identification of the causal agent of Botryosphaeria stem canker in Ethiopian Eucalyptus plantations south African. Journal of Botany. 2004;**70**(2): 241-248 ISSN 0254–6299

[7] Chen SE, Gryzenhout M, Roux J, Xie YJ, Wingfield MJ, Zhou XD. Identification and pathogenicity of Chrysoporthe cubensis on Eucalyptus and Syzygium spp. in SouthChina. Plant Disease. 2010;**94**:1143-1150. DOI: 10.1094/PDIS-94-9-1143

[8] Rosa S, Soelistijono J. Development of pink disease on various ages of *Acacia crassicarpa* CUNN EX. BENTH. Jurnal Hama Dan Penyakit Tumbuhan Tropika: Journal of Tropical Plant Pests and Diseases. 2020;**20**(2):116-122. DOI: 10.23960/j.hptt.220116-122

[9] Iin A, Djumali M. Diseases of shoot, stem and root on *Eucalyptus pellita* F. Muell. at Sebulu, East Kalimantan. Advances in Biological Sciences Research. In Proceedings. Joint Symposium on Tropical Studies (JSTS-19). 2021;**11**:186-192. DOI: 10.2991/absr.k.210408.031

[10] Li GQ, Liu FF, Li JQ, Liu QL, Chen SF. Botryosphaeriaceae from Eucalyptus plantations and adjacent plants in China. Persoonia. 2018;**40**: 63-95. DOI: 10.3767/persoonia.2018.40.03

[11] Rasdianah D, Paridah M, Adlin SMR, Lee SH, Edi SB, Zaidon A, et al. Influence of Chrysoporthe deuterocubensis Canker Disease on the Physical and Mechanical Properties of Eucalyptus urograndis. Forests. 2021; **12**(5):639. DOI: DOI. 10.3390/f12050639

[12] Chenab SF, Pavlica D, Rouxa J, Slippersc B, Xieb XJ, Wingfielda MJ, Zhouab XD. Characterization of Botryosphaeriaceae from plantation-grown Eucalyptus species in South China. Plant Pathology. 2011;**60**(4): 186-192. DOI: 10.1111/j.1365-3059.2011.02431.x

[13] Wingfield MJ, Roux J, Slippers B, et al. Established and new technologies reduce increasing pest and pathogen threats to eucalypt plantations. Forest Ecology and Management. 2013;**301**: 35-42

[14] Romero PIA, Ocampos CG, Carpio C, Toro VS, Ferreira AFTA. Mizubuti ESG first report of *Ralstonia solanacearum* causing bacterial wilt of Eucalyptus in Ecuador. Plant Disease. 2020:31. DOI: 10.1094/PDIS-11-19-2516-PDN

[15] Ragil SBI, Kara MB, Nur H, Ito S, Fiani A, Anto R, et al. Incidence and spatial analysis of root rot of Acacia mangium in Indonesia. Journal of Tropical Forest Science. 2006;**18**(3)

[16] Lee SS. Observations on the successes and failures of Acacia plantations in Sabah and Sarawak and the way forward. Journal of Tropical Forest Science. 2018;**30**(Anniversary Issue):468-475. DOI: 10.26525/jtfs2019.30.5.468475

[17] Arisman H, Hardiyanto EB. Acacia mangium – A historical perspective on its cultivation. Pp 11–15. In: Potter K, Rimbawanto A, Beadle C, editors. Heart Rot and Root Rot in Tropical Acacia Plantations. Proceedings. ACIAR Proceedings. Canberra: Australian Centre for International Agricultural Research (ACIAR); 2006. p. 124

[18] Mohammed C, Beadle C, Francis A, et al. Management of Fungal Root rot in Plantation Acacias in Indonesia. Final Report for ACIAR Project [FST/2003/048], Australian Centre for International Agricultural Research (ACIAR). Canberra: Environmental Science; 2012. Available from: https://www.aciar.gov.au/node/10706

[19] Jeremy TB, Sri S, Arif N, Heru I, Marthin T, Muhammad Y, et al. Screening clonally replicated Acacia mangium breeding populations for tolerance to Ceratocystis canker and wilt disease. Tree Genetics & Genomes. 2022;**18**:17. DOI: 10.1007/s11295-022-01548-9

[20] Brawner JB, Yani J, Lapammu R, David AB, Wingfield M. Evaluating the inheritance of Ceratocystis acaciivora symptom expression in a diverse *Acacia mangium* breeding population. Southern forests a. Journal of Forest Science. 2015;77 (1):83-90. DOI: 10.2989/20702620.2015.1007412

[21] Sadanandan EKN, Christopher EH, Nguyen DK. Acacia plantations in Vietnam: Research and knowledge application to secure a sustainable future. Southern Forests: A Journal of Forest Science. 2014;77:1-10. DOI: 10.2989/20702620.2014.999301

[22] Sri R, Sapto I, Tito A. Spatial and temporal distribution of pink fungus Corticium Salmonicolor in Lanete, jackfruit and Indian prickly ash plantation. International Journal of Agriculture, Forestry and Plantation. 2021;**11**:165-173

[23] Francis A, Beadle C, Puspitasari D, Irianto R, Agustini L, Rimbawanto A, et. al. Disease progression in plantations of Acacia mangium affected by red root rot (Ganoderma philippii). Forest Pathology. 2014;**44**(6):447-459. 10.1111/efp.12141

[24] Old KM, Hood IA, Yuan ZQ. Diseases of tropical acacias in northern Queensland. In: Old KM, See LS, Sharma JK, editors. Proceedings of an International Workshop Diseases of Tropical Acacias. Subanjeriji, South Sumatra. Jakarta, Indonesia: Center for International Forestry Research; 1997. pp. 1-22

[25] Aylward JF, Dreyer LL, Wingfield MJ. Teratosphaeria stem

canker of Eucalyptus: Two pathogens, one devastating disease. Molecular Plant Pathology. 2019;**20**(1):8-19. DOI: 10.1111/mpp.12758

[26] Shuai FC, Marieka G, Jolanda R, Xie YJ, Michael JW, Xudong Z. Identification and Pathogenicity of Chrysoporthe cubensis on Eucalyptus and Syzygium spp. in South China. Plant Disease. 2010;**94**(9):1143-1150. DOI: 10.1094/PDIS-94-9-1143

[27] Geoffrey SP, Marieka G, O'D C, Andre D, Michael JW. The Eucalyptus canker pathogen Chrysoporthe cubensis discovered in eastern Australia. Australasian Plant Pathology. 2010; **39**(4):343-349. DOI: 10.1071/AP10004

[28] Sri S, Na'iem M, Hardiyanto EB, Sapto I. Breeding strategy of Acacia hybrid (Acacia mangium × a. auriculiformis) to increase Forest plantation productivity in Indonesia. Journal of Tropical Forest Management. 2013;**19**(2):128-137. DOI: 10.7226/jtfm.19.2.128

[29] Thu PQ, Griffiths MW, Pegg GS, McDonald JM, Wylie FR, King J, et al. Healthy Plantations: A Field Guide to Pests and Pathogens of Acacia, Eucalyptus and Pinus in Vietnam. Queensland, Australia: Department of Employment, Economic Development and Innovation; 2010. p. 124

[30] Chin FH. Pink disease - its incidence and economic importance in Sarawak, Malaysia. In: Proceedings Paper Malaysian Plant Protection Society. 2013 http://agris.upm.edu.my:8080/dspace/handle/0/5021

[31] Carstensen GD, Venter SN, Wingfield MJ, Coutinho TA. Two Ralstonia species associated with bacterial wilt of Eucalyptus. Plant Pathology. 2017;**66**:393-403. DOI: 10.1111/ppa.12577

[32] Rahayu S, See L, Shukor N, Saleh N. Environmental factors related to gall rust disease development on Falcataria moluccana (miq.) Barneby & J. W. Grimes at Brumas estate, Tawau, Sabah, Malaysia. Applied Ecology and Environmental Research. 2018;**16**(6):7485-7499. DOI: 10.15666/AEER/1606_74857499

[33] Richmund P, Lydia ET, Lydia, Oliva C, et al. Gall rust disease incidence of Falcata (Paraserianthes falcataria (L.) Nielsen) in Falcata – Based agroforestry systems in Misamis Oriental, Philippines. 2020 IOP Conference Series: Earth and Environmental Science. 2019; **449**:012035. DOI: 10.1088/1755-1315/449/1/012035

[34] Doungsa-ard C, McTaggart AR, Geering ADW, et al. *Uromycladium falcatarium* sp. nov., the cause of gall rust on *Paraserianthes falcataria* in south-East Asia. Australasian. Plant Pathology. 2015;**44**:25-30. DOI: 10.1007/s13313-014-0301-z

[35] Siregar BA, Giyanto, Hidayat SH, Siregar IZ, Tjahjono B. Epidemiology of bacterial wilt disease on Eucalyptus pellita F. Muell. in Indonesia. Proceeding: Southeast Asia Plant Protection Conference 2019 IOP Conf. Series: Earth and Environmental Science. 2020;**468**:012033. DOI: 10.1088/1755-1315/468/1/012033

[36] Bayo AS, Giyanto, Sri HH, Iskandar ZS, Budi T. Diversity of Ralstonia pseudosolanacearum, the causal agent of bacterial wilt on Eucalyptus pellitain Indonesia. Biodiversitas. 2021;**22**(6):2085-4722. DOI: 10.13057/biodiv/d220664

[37] Jun Y, Guanghui D, Shurong W. China's National Monitoring Program on

ecological functions of forests: An analysis of the protocol and initial results. Forests. 2015;**6**:809-826. DOI: 10.3390/f6030809

[38] Wingfield MJ, Slippers B, Hurley BP, Coutinho TA, Wingfield BD, Roux J. Eucalypt pests and diseases: Growing threats to plantation productivity. Southern Forests. 2008;**70**:139-144. DOI: 10.2989/SOUTH.FOR.2008.70.2.9.537

[39] Dahali R, Tahir MD, Roseley ASM, Hua LS, Bakar ES, Ashaari Z, et al. Influence of Chrysoporthe deuterocubensis canker disease on the physical and mechanical properties of Eucalyptus urograndis. Forests. 2021;**12**:639. DOI: 10.3390/f12050639

[40] Megat Mohamed Nazir MN, Terhem R, Norhisham AR, Mohd Razali S, Meder R. Early monitoring of health status of plantation-grown Eucalyptus pellita at large spatial scale via visible Spectrum imaging of canopy foliage using unmanned aerial vehicles. Forests. 2021;**12**:1393. DOI: 10.3390/f12101393

[41] Rauf MRBA, McTaggart AR, Marincowitz S, et al. Pathogenicity of Chrysoporthe deuterocubensis and Myrtoporthe bodenii gen. Et sp. nov. on Eucalyptus in Sabah, Malaysia. Australasian Plant Pathology. 2020;**49**:53-64. DOI: 10.1007/s13313-019-00676-6

Chapter 4

Main Pests and Diseases in Tropical Forest Species in Nursery

Ángel Sol Sánchez, Gloria Isela Hernández Melchor and Facundo Sánchez Gutiérrez

Abstract

This paper presents the monitoring of pests and diseases in nurseries of 10 tropical forest species used in the reforestation of disturbed areas. The work was carried out in a rustic nursery established in Cardenas, Tabasco, under cocoa shade. The objective was to evaluate the presence of pests or diseases in the nursery under natural conditions. Pests and diseases appeared from the seedling stage in germination beds to the adult stage. The fungus *Fusarium* was the most aggressive causal agent that caused the death of seedling in the germination beds, as well as *Curvularia lanata* that massively affected *Tabebuia rosea* plants. Likewise, *T. rosea* is one of the species with the most reported pests, as well as *Lantana camara*. On the contrary, *Hamelia patens* Jacq (Coralillo) did not registered important pests during the monitoring in the nursery.

Keywords: tropical foest species, pests, diseases, nursery, plants

1. Introduction

In the tropics, the diversity of plant species is very high, although the species richness is not so great because they share the habitat with the other biological forms of that environment, such as shrubs, vines, lianas, parasitic epiphytes and palms.

In Mexico, for a long time precious woods such as cedar (*Cedrela odorala* L.), mahogany (*Swietenia macrophylla* King.), spring (*Cybistab Donnell-Smith*) and red (*Haematoxylum campechianum* L.) were exploited to take them to other continents for various cabinetry work, to the extent that the ecosystems of the Mexican tropics were impoverished by an extraction of eugenic type, that is, taking advantage of the best quality trees with straight stems and larger diameters.

This extraction in some cases has caused the disappearance of habitats due to extractive recurrence, thus favoring that the trees do not reach their physiological maturity, although their reproductive maturity does.

On the other hand, the establishment of exotic forest plantations such as teak (*Tectona grandis* L.f.), rubber (*Hevea brasiliensis* [Willd. ex A. Juss.] Müll. Arg.), melina (*Gmelina arborea* Roxb), pine (*Pinus caribaea* Var. Hondurensis [Sénécl] W.H.G.) or eucalyptus (*Eucalyptus* sp), Oil palm (*Elaeis guineensis* Jacq.) by large corporations, has caused the reduction of the surface of the original vegetation, because when establishing these plantations all the vegetation present is eliminate, and far from supporting the local population, it exterminates the environments of the area, leaving only pockets of impoverished land to restore part of the lost biological diversity.

In this sense, forest nurseries of local species are created in order to provide seedlings to recover the strongholds of impoverished areas, potential environments or even include other environments for conservation.

For this purpose, the availability of germplasm is an activity of transcendental importance in which the possible progenitors are located, the period of flowering and fruiting is expected and later the seed is collected. This seed is taken to the nursery, processed, sown and the emergence of the seedlings is expected. At this stage, the management and care of the seedlings is of utmost importance because they are more susceptible to attack by pests and/or disease of any pest or illness.

Under this concept, in order to generate information about the main pests of some forest species used in reforestation in a mixed way, ten germination and growth beds of various local forest species were established. It is possible that due to the difference in habitat conditions between the site where they grow naturally and the nursery conditions where they were produced, some pests and diseases were present both in germination beds and growth beds and that are reported in this work.

Currently, it is a common practice to produce forest plants in a nursery and when the seed does not come from certified seed banks, these generally already bring the pest from the harvest field in a certain percentage, which are not suitable for cultivation.

When the percentage of impurities has been eliminated, the resulting seed is sown in bags, tubes, trays or germination beds depending on the size of the seed and wait for them to germinate. During the process of germination and growth of seedlings some diseases occur.

1.1 Damping off *Rizotocnia*

This is a type of opportunistic fungus that occurs when excess moisture in the site is above what the plant can dispose of. This excess moisture ends up rotting the stem of the seedling. They generally call it a dryer and it is common in various species of plants in the region such as cocoite, cedar, bojon, guanacaste, to name a few species and generally what is presented is the mycelium since they rarely form structures of sexual reproduction.

This fungus is very damaging as it forms resistant structures called interconnected Sclerotia that infect the roots of newly emerging plants and can rapidly spread over much of the germination bed.

Its presence is greater in alkaline soils with poor drainage, so generally the substrate must be fumigated with methyl bromide before sowing, covered with plastic and uncovered after two days so that there is aeration, and the seed is sown after 72 hours of aeration.

Another very common fungus is Fusarium sp., and it occurs when the seed generally does not go through an antifungal treatment before sowing, so the fungus Fusarium oxysporum attacks the external and even internal structures of the seeds. The newly emerged seedlings are covered with a white mycelium that affects the new organs of the plant, which is controlled using thiabenzadol at a dose of 1 g per l of water.

2. Germoplasm collection

During the year 2020, germplasm of various forest species or species of interest for biological diversity was collected, this in order to produce plants in germination beds, these species were: *Cedrela odorata* L., *Bursera simaruba* (L.) Sarg, *Cordia megalantha* S. F. Blake, *Hamelia patens* Jacq., *Pimenta dioica* L. (Merr), *Tabebuia rosea* (Bertol) Dc, *Lantana camara* L., *Sterculia apetala* (Jacq.) H. Karsten, *Roystonea regia* (Kunth.) O. F. Cook, and *Chrysobalanus icaco* (L.) L.

The collection of fruits was carried out by various methods such as collection from the ground, collection with a pole and hand scissors, manual selection of the fruits in the branches, specifically for the fruits of the royal palm, it was necessary to climb the palm using a team of ropes to electricians. These fruits were transported in jute sacks to the nursery and the benefit was made there.

Since the palm fruits have a very hard test, they were kept in the sack for a week, then they were washed and then physical scarification was carried out to accelerate the germination.

The seeds of all species were scarified to ensure germination. Specifically, for *H. patens* Jacq (Coralillo), because seeds were collected twice and germination did not occur, cuttings of approximately 20 cm in length were collected. These were placed in a bucket with a prepared solution of rootstock in a proportion of 15 g per 5 l of water. Here the basal part of the cutting was submerged for 5 minutes and then it was planted in the bags of soil.

The substrate used for the germination beds was grit enriched with filter cake, which is a residue obtained in the cane juice clarification process, which includes earthy materials and organic impurities [1].

2.1 *Cedrela odorata* L., (red cedar)

The planted cedar germplasm was quality seed, selected and of regular size, and reached its total germination at 18 days, this presented some problems in the germination beds, mainly due to excess water and lack of drainage, which generated problems of germination and presence of *Fusarium* and *Pythium*.

In the already bagged plants, the Meliaceae borer *Hypsiphylla* grandella was present, and caused problems in 6% of the plants, which were lost due to the bored stem [2]. To avoid greater severity of the damage, sprays were applied both to the plants in bags and to the adjacent beds that were in the process of growth.

Worldwide the main pests of cedar are insects of the genera *Anacrusis* sp. (Tortricidae), *Antaeotricha ribbel* (Stenomidae), *Hypsyphylla grandella*, as well as ants of the *Atta* genus. For Costa Rica, the presence of 15 pests in cedar and two diseases caused by Fusarium Sp and *Phyllachora balansae*, in addition to the species *Orthogeomys heterodus* was reported [3].

The presence of a cedar leaf-roller worm was also recorded, the presence of this pest was recurrent at least four times in six months while the plants were in the nursery. In the substrate, a large number of blind chicken caterpillars (*Phyllopaga* sp), which are beetle larvae, were recorded, but they did not cause any damage to the cedar plants.

In some plants the mealybug (*Mastigimas* sp) appeared, covering the entire plant, which caused its weakening and death due to the overpopulation of the mealybug, these plants were eliminated. For Mexico also was reported nine species of pests for cedar [4]. The insects reported for Costa Rica and México are listed in **Tables 1** and **2** [3, 4].

Insect	Family	Affected part
Anacrusis sp	Tortricidae	Foliage
Antaeotricha ribbel	Stenomidae	Foliage
Apatelodes sp	Apatelodidae	Foliage
Atta sp	Formicidae	Foliage
Hypsypylla grandella	Pyralidae	Shoots, seeds
Mastigimas sp	Psillidae	Foliage
Natad sp	Limacodide	Foliage
Phyllocnistes meliacella	Gracilariidae	Foliage
Sematoneura atrovenosella	Piralidae	Seeds
Sematoneura grijmani	Piralidae	Seeds
Taeniopoda sp	Romaleidae	Foliage
Thecla cupentus	Lycaenidae	Foliage

Table 1.
Insects reported for red cedar in Costa Rica [3].

Specie	Common name	Distribution in Mexico
Cicadellidae	sharpshooter	Wide distribution
	mealy louse	Campeche, Jalisco, Oaxaca, Puebla Quintana Roo, Tabasco, Tamaulipas, Veracruz and Yucatán
Atta cephalotes	leafcutter ant	Campeche, Chiapas, Oaxaca, Quintana Roo, Tabasco and Veracruz
Atta mexicana	leafcutter ant	Guerrero, Jalisco, Nayarit Tamaulipas and Veracruz
Hypsypylla grandella	borer	Wide distribution
Chrysobothris	borer	Wide distribution
Nasutitemis	termites	Wide distribution
Phyllophaga	blind chicken	Wide distribution
Tetranychis	red spider	Hidalgo, Oaxaca, Puebla and Veracruz

Table 2.
Cedar pest insects reported for México [4].

2.2 *T. rosea Bertol* dc. (Macuilis)

Despite the fact that macuilis is one of the hardest species in the state and, in turn, one of the hardest woods, it presents a great diversity of pests and diseases in the nursery. One of the main ones that appeared was the Curvularia lanata fungus, which quickly withered a large number of plants, for which it was necessary to remove the burned leaves, apply fungicide and promote ventilation to the growth beds.

Likewise, the so-called spring rust of *Prospodium perornatum* appeared sporadically on some of the approximately five-month-old *Tabebuia rosea* Plants.

The affected part showed its spermogonia as reddish dots on deep green swollen areas; especially in the upper part of the pedicel of the leaf forming an arch. This situation caused the infected leaves to break, for this reason the plant was eliminated.

A recurring pest was the *Eulepte gastralis* worm, which was present at different stages and which caused damage to the leaves causing their skeletonization, whose protection is a mesh of silk and excrement. A large population of blind hen (*Phyllophaga* sp) was recorded in the substrate, however, since its presence was in the rainy season, it did not cause damage to the macuilis plants.

For macuilis, it has been reported that it is susceptible to attack by nematodes of the genus Meloidogyne incognita, which directly attack the root part [5]. *Eulepe Gastralis* (Lepidoptera: Pyralidade) in the nursery presents a relationship between density and stratum so that the plant is affected. Other authors report a wide diversity of pests for Mexico and the Roca coast **Table 3** [3, 4].

Pest	Family	Stage	Type of affectation
Modoryx oielus Clarke	Lepidoptera: Sphingidae	cobra worm a	eats leaves
Megalostis anacoreta L.	Coleoptera. Chysomelidae	greater hog	gnaws buds
Megistops sp	Coleoptera. Chysomelidae	Oak manakin	gnaws buds
Oiketicus kirbyi (Lands-Guilding)	Lepidoptera: Psychidae	básquet worm	eats leaves
Oncideres teddellata (Thoms)	Coleoptera:cerambycidae	spotted harlequin	pierces stems
Phobetron hipparchia Cramer	Lepidoptera Limacodidae	spider worm	eats leaves
Prosarthia teretrirostris Brunn	Ortopthera: proscopidae	flat shackle	eats leaves
Protaleura tabebuiae Dozier	Homoptera cicadellidae	leafhoppers	sucks sap
Rabdotalebra sp	Homoptera cicadellidae	leafhoppers	Sucks sap
Steirastoma hishistrionicum thams	Coleoptera:cerambycidae	harlequin engrabado	Pierce stems
Tenuipalpus sp	Acari: Tenuipalpidae	Flat mite	Sucks sap
Urodera sp	Coleoptera. Chysomelidae	humpbacked beetle	Gnaws leaves
Zygogramma cognata Stal	Coleoptera. Chysomelidae	Engraved morrocoyita	Punches sheets
Amphicerus cornutos (Pallas)	Coleoptera: Bostrichidae	Capuccino	Pierce stems
Archegozetes sp.	Oribatida: epilohmanidae	painted mite	Terminals
Atta sp	Hymenoptera:Formycidae	leafcutter ant	cut leaves
Automeris junonia	Lepidoptera. Saturniidae	spiny worm	eat leaves
Automeris sp	Lepidoptera. Saturniidae	spiny worm	eat leaves
Compsus sp	Coleoptera: Curculionidae	Little	eat leaves
Cryptotermes brevis (W.)	Isoptera: Kalotermitidae	termite	Pierce the base of the stem
Diabrotica sp	Coleoptera:Chrysomelydae	Morrocoyita	Log leavess
Dikraneura sp	Homoptera:Cicadelidae	leafhoppers	Sucks sap, hojas
Edessa leucogramma Perty	Hemiptera:Pentatomidae	Guayacan jackdaw	Sucks sap

Pest	Family	Stage	Type of affectation
Eriophyes sp.	Acarina:Eriophydae	roller	gills leaves
Eulepte gastralis guanee	Lepidotera:Pyralidae	sheet gluer	eat leaves
Gastrotrips sp.	Thysanoptera:Phlaeotripidae	black trips	Micetophage
Halisidota sp	Lepidoptera:Arctiidae	silly chapola	eats leaves
Hemeroplanes Parce F.	Lepidoptera:Sphingidae	cobra worm	eats leaves
Hyphotenemus sp	Coleptera: Solytinae	scolithid	Pierce stems
Lagochirurs araneiformis LInn	Coleoptera Cerambicidae	painted Brown harlequin	Pierce stems and brances ramas
Lepydomys sp	Lepidoptera: Pyralidae	oak borer	pierces terminals

Table 3.
Pests reported for Tabebuia rosea [3, 4].

2.3 *R. regia* (Kunth) O.F. Cook (Royal Palm)

This palm was very susceptible to the Fusarium fungus during the nursery stage, mainly in the pinnae of the fronds. This fungus was present after the rainy season. Causing necrosis on the sides of the pinnae and in the terminal part of the frond, which caused dieback of an entire batch of these palms in the nursery. The fungicide Captan was applied repeatedly with few positive results, although depending on the severity of the damage, the application response could be any other fungicide [6].

2.4 *L. camara* L.

Being a fast-growing species, it is offered as an attraction for various nursery pests. During the evaluation period in the nursery, this plant grew and flourished and in turn attracted pests such as *Tetranychus urticae* (red spider), some type of aphid and some unidentified caterpillar. Disorganized plant growth provided safe haven for these pests, however, they were not considered severe pests as they did not cause problems to the plants.

This species as an aggressive weed for livestock capable of invading agricultural areas if it is not managed properly [7]. In this regard, were reported the presence of 118 pest species, some in all stages of development [8]. In the state of Tabasco, this species is found in abandoned agricultural areas or is part of the ruderal areas, but because the agricultural surface per producer is not very extensive, every two years the land for crops is cleared, for this reason it does not extend aggressively. Some authors delimit a wide diversity of 29 species of common pests for this species, some of the most frequent are *Tetranychus urticae* spider or red spider: this spider weaves its web on the underside of the leaves, so it cannot be seen with the naked eye and feeds on the sap of the leaves [9]. This species of pest was very common in the six months of the nursery and its presence was frequent, its elimination was with chemical products.

Cochineal: The presence of this pest in Lantana plants is common [10], however; in the nursery during the evaluation period it did not appear because the cochineal occurs mostly in sunny areas and the nursery maintains an average of 80% shade.

Bemicia tabaci (whitefly) appeared as a recurring pest in the nursery, its presence was manifested since the seedlings were small; This species lays its eggs on the underside of the leaves, so the larvae feed on the sap when they emerge. The occasions that arose caused the deaths of secondary branches of the lantana plants, most of the time the plants were pruned.

Hemiptera (aphids) were present during the entire evaluation period that they were monitored. This pest is detected in the mornings on sunny days. These feed on the plants by sucking the sap, the highest concentration of aphids were located in the tender shoots or apical buds. Among the main symptoms are very sticky leaves due to the molasses excreted by these insects. His treatment was with chemicals.

Moths, a type of grayish brown moth called *Neogalea sunia* was present, it also lays its eggs on the plant and when they hatch and grow, they feed on the foliage of the plant and then become its pollinator. It is recognized because the larvae when feeding leave evidence of eating in the form of circles, their eradication is at that stage, as they are very voracious due to their color, they are easily confused with the stem or leaves of the plant and are easily lost.

Some of the diseases reported for Lantana are *Alternaria*, which manifests itself with brown spots on the leaves, in addition to Fusarium, which was very aggressive in the nursery, since it caused the decay of the plant, since it was necessary to apply antifungals, avoid irrigation and encourage the entry of sunlight.

2.5 *B. simaruba* (L.) Sarg

This species, despite being a rough species and well adapted to tropical conditions, presented some species of pest insects; however, the damages were not severe and did not represent damage per se to the plants in the nursery.

It was detected that the insects do not use the mulatto plants as a habitat, rather as occasional visitors, that when the plant is monitored, they quickly disappear by flying.

Some especies of pests for palo mulato, such as the beetles *Xyleborus* spp. and *Platypus* spp., as pests of green wood, *Lagochirus araneiformis* L. and other insects bore into the bark and wood, they also feed on wood and living seedling trees and Lyctus spp. attacks dry wood like termites [11].

On the other hand, despite the toughness of the plant, the *Fusarium* fungus caused serious problems in entire batches of plants because the substrate used retained a lot of moisture, also when the plant was transplanted from the germination bed to plastic containers., there was high mortality caused by Fusarium. This species tolerates abundant irrigation well, but with good filtration and 60% on shade.

2.6 *Cordia megalantha* S. F. Blake (Candelestick)

This species, at the level of plants in germination beds and bagged, did not present serious problems, occasionally some circular insect bites were observed in the dry season of the year.

However, in sowing the germination bed in the bag, mortality was very high due to the presence of Fusarium, which also occurred in royal palm and other species, due to stem, rot.

It was observed that its growth is limited when shade exceeds 65% and in turn favors the arrival of occasional insects that feed on new shoots. Likewise, when the incidence of sunlight exceeds 60%, the leaves turn yellowish and slow down their growth.

2.7 *S. apetala* (Jacq.) H. Karsten

In the nursery stage, important pests were observed, except that the lack of irrigation caused the plants to decay and lose their turgor. The larval phase of Micromartinia mnemusalis has been reported as a pest of S. apetala for Costa Rica [10].

The adult stage corresponds to various moths considered as pests for various species of commercial interest, however, during the monitoring phase in the nursery it was not present. This plant species was one of the least affected by pests or diseases.

2.8 *Pimienta dioica* L.

Pepper is a species of great interest in food and for the extraction of various oils. In some states of the republic it is not cultivated in plantations, but as shade species of other crops, such is the case of cocoa in Tabasco, or borders in pastures. Although it does not receive management, its production is considered good.

Pepper seedling age, it generally does not present pests or diseases, but it is very demanding in relation to well-drained soils and sufficient humidity, it grows well in any type of soil, from very fertile to very poor such as arenosols and savannah soils.

In the nursery and in newly established plantations, this fungus has the greatest impact. Infections present as isolated necrotic lesions that vary in size from light brown to almost black in color and may cause plant death. When the damage is severe, a yellow powder is generated on the underside of the leaf and when the inflorescences are affected, they turn black and finally die. Control could be carried out by spraying an antifungal [12].

2.9 *H. patens* Jacq (Coralillo)

It is a bushy species, abundant in the region, with a wide distribution from Florida to Argentina, and it is difficult to propagate by seeds. In the nursery, the seeds take up to eight months to germinate and their initial growth is too slow, so it was decided to reproduce them by cuttings. The most appropriate thickness for reproduction was one cm in diameter on average.

The cut to chop the stakes must be clean and each stake must not exceed 20 cm in length. When the cut is not clean, the bark is mistreated and that is where the rot caused by *Fusarium* begins.

This species is not of commercial importance, but it is of local use. In cultivated areas, its growth is allowed, as it attracts pests in agricultural crops. It is attracting pollinating bees, hummingbirds and butterflies. There were no pests in the nursery during the evaluation period.

2.10 *C. icaco* (L.) L. (Icaco)

For the establishment of the germination beds of Icaco, the seeds were first collected from the ground and later the fruits of the tree were collected.

The seeds collected from the ground were perforated by some unidentified insect, therefore, there was no germination.

Subsequently, ripe fruits were collected and put to ferment for three days in a plastic bag with the mouth closed, then the pulp was removed, the seed was washed and a physical scarification was performed, under this method germination was obtained at 28 days.

During the germination phase, there were problems with fungi that caused the death of some plants. After transplanting to the bags, the *Fusarium* fungus was also present, causing the death of some plants, and the *Curvularia* sp. fungus was frequent, causing the death of some of the terminal leaves of the plants. For this species it was reported the presence of the diptera *Anastrepha* (flies), as well as *Ceratitis capitata* and *thephytophagous* mite and *Oligonychus bagdasariani* [13].

Other authors report the presence of the Red Spider Tetranychus urticae, which causes yellowing of the leaves at the site of the bites and when it comes to attacking fruits, they acquire a dirty color due to the dark spots of the bites.

3. Conclusion

The presence of pests and diseases in the seedling stage was common. In the germination beds, Fusarium behaved recurrently in most of the germination beds, mainly in *C. odorata* and *R. regia*. The most specific pests *were Hypsiphylla grandella* for *cedrela odorata*, the fungus *Curvularia lanata* for T. rosea, as well as *Prospodium perornatum* for the same species.

Excess shade, continuous irrigation, residual moisture and evapotranspiration were key factors of importance for the development of fungi almost during the six months of monitoring of the plants in the nursery, although shade was removed from the nursery trees twice this year. it reappeared in a short time.

The pests entered the nursery in the area where the nursery adjoins herbaceous and shrubby vegetation, that is, the natural habitat of these organisms is that type of environment. It was recorded that the more the plants grow, the greater the possibility of shelter for pests.

The reason for including shrubby species such as Icaco, or herbaceous species such as lantana and Hamelia, is that in disturbed areas, herbaceous plants are the pioneers of secondary succession, and when reforestation or restoration areas are established, herbaceous species are very important for maintaining soil cover and protection. In addition, they are very important in attracting beneficial insects for Pollination.

Due to the abuse of agrochemicals for various uses, biological diversity (Flora and fauna) has been greatly reduced, so it is necessary to restore ecological conditions and habitats to generate a status of ecological balance at various scales in time. Hence, it is very important to know the pests or diseases that could affect the species used in habitat restoration.

In any environment that has been altered, pests settle in the islands of vegetation present, so their eradication is almost impossible due to their qualities of adaptation to the changing environment.

In this way, the species that are planted will have different functions over time in which the ecosystem normalizes and the insect pests adapt to that environment.

Author details

Ángel Sol Sánchez[1], Gloria Isela Hernández Melchor[2]*
and Facundo Sánchez Gutiérrez[2]

1 Postgraduate College, Tabasco Campus, Tabasco, México

2 Mayan College of Higher Studies, Catazaja Campus, Autonomous University of Chiapas, Catazaja Chis, México

*Address all correspondence to: isela7827@gmail.com

References

[1] Hernández MGI, Salgado GS, LDJ P, Del ELC L, Castelán EM, Ruiz RO. Vinaza y composta de cachaza como fuente de nutrientes en caña de azúcar en un gleysol mólico de Chiapas, Mexico. Interciencia. 2008;**33**(11):855-860. Available from: http://ve.scielo.org/scielo.php?script=sci_arttext&pid=S0378-8442008001100016&lng=es&tlng=es

[2] Ramírez-García C, Vera-Castillo G, Carrillo-Anzures F, Octavio Salvador Magaña Y, Torres OS. El cedro rojo (*Cedrela odorata* L.) como alternativa de reconversión de terrenos agrícolas en el sur de Tamaulipas. Rev. Tec. Mex. 2008;**34**:243-250

[3] Arguedas-Gamboa M. Plagas y enfermedades forestales en Costa Rica. vol. **4**(11-12). Kurú: Instituto Tecnológico de Costa Rica, Revista Forestal de Costa Rica; 2007. p. 77

[4] Cibrian TD. Manual para la identificación y manejo de plagas en plantaciones forestales comerciales. Comisión Nacional Forestal. Universidad Autónoma Chapingo; 2013. p. 238

[5] Hernández R. F Y Briceño VA. Factores bióticos y abióticos que favorecen el desarrollo de Eulepte gastralis, en vivero de apamate. Universidad de Los Andes, Facultad de Ciencias Forestales y Ambientales. Centro de Estudios Forestales y Ambientales de Postgrado. Mérida-Venezuela Rev. Forest. Venez. 1998;**42**(2):157-166

[6] Tahir BR, Shakeel Q, Muhammad R, Rashid I. Fungal diseases of royal palm (Roystonea regia). In: Etiology and Integrated Management of Economically Fungal Diseases of Ornamental Palms. Springer Nature; 2020. pp. 222-235

[7] Matienzo Y, Ramos B, Rijo E. Revisión bibliográfica sobre Lantana camara L. una amenaza para la ganadería. Fitosanidad. vol. 7(4). Instituto de Investigaciones de Sanidad Vegetal, La habana; 2003

[8] Palmer WA, Pullen KR. The phytophagous arthropods associated with Lantana camara, L. irsuta, L. urticifolia and L. urticoides (Verbenaceae) in North America. Biological Control. 1995;**5**:54-72

[9] Vargas BB, Yoannia Gretel PBY, Fajardo R, Puertas LAA, Rizó MM. Diversidad de Insectos asociada a Lantana Camara L. (rompe camisa) en localidades agrícolas de Santiago de Cuba, Cuba. Investigación y Saberes. 2015;**Vol, IV, No. 1**(2015):17-28

[10] Cantillana E. Micromartinia mnemusalis. 2019. Available from: https://www.acguanacaste.ac.cr/paginas-de-especies/insectos/296-crambidae/4633-i-micromartinia-mnemusalis-i-crambidae

[11] Francis JK. Bursera Simaruba (L.) Sarg Burseracea. Bursera Family. 1990. Available from: chrome-extension://efaidnbmnnnibpcajpcglclefindmkaj/https://www.fs.fed.us/global/iitf/pubs/sm_iitf035%20%20(5).pdf, p. 5

[12] Martínez PM, Hernández G, Martínez G. La Pimineta Gorda en Mexico (Pimenta dioica L. Merril): Avances y retos en la gestion de la iNnovacion. Coleccion Tropico Humedo. Universdad Autonoma Chapingo; 2013. p. 73

[13] Quirós de GM, Viloria Z. Tetranychus Urticae KOCH y Oligonychus Bagdasariani Baker y Pritchard, (ACARI: TETRANYCHIDAE) Acaros. 1991. En línea: https://www.intagri.com/articulos/frutales/el-caco-como-cultivo-potencial

Chapter 5

Perspective Chapter: Microorganisms and Their Relationship with Tree Health

Rodrigo F. Ramos, Lisiane Sobucki, Estéfany Pawlowski, Janaina S. Sarzi, Jessica E. Rabuske, Lucas G. Savian, Tiago E. Kaspary and Cristiano Bellé

Abstract

The health of plants depends on numerous environmental factors. All plants, including trees, live in close relationship with microorganisms. Plants harbor microbial communities in above- and below-ground tissues, where plant-associated microbial communities are influenced by environmental conditions and host genotype. The microbiome of trees is composed of mutualistic, commensal, and pathogenic microorganisms. Mutualistic microorganisms can help trees obtain nutrients (e.g., phosphorus and nitrogen) and defend against plant pathogens. Ecological interactions between different microbial groups directly influence host health, and endophytic microorganisms can inhibit pathogen growth or induce the expression of genes related to tree defense against these adverse organisms. Hence, understanding host-microbiome-environment interactions are crucial for modulating tree health.

Keywords: plant holobiont, microbial ecology, bacteria, ectomycorrhizas, arbuscular mycorrhizae

1. Introduction

Trees, including those outside of forests (i.e., orchards, gardens, trees in urban environments), are terrestrial plants that provide several ecosystem services essential to life and the economy. Trees contribute to carbon sequestration, animal and insect biodiversity maintenance, lignocellulosic biomass production for the industry, food supply (i.e., fruits and seeds), tourism (i.e., forest nature trails), air filtration in urban areas, and wood production for construction. Given their importance and the benefits trees provide, understanding the factors related to their health is pivotal for proper human intervention in forests, woodlands, and orchards.

Tree health depends on numerous factors, and it is now understood that their health, as well as other multicellular organisms, is closely related to the structure of the microbial community of the holobiont (the host and its associated microorganisms) [1]. The theory of plant holobiont seeks to shed more light on the plant as a meta-organism, in which plant growth, health, and productivity are closely related to the composition and functions of the microbiota inhabiting different niches in the plant [2]. Within this perspective, trees are not autonomous entities, and their health can be considered a function of the plant microbiota (set of all associated microorganisms) or the plant microbiome (set of all associated microbial genomes) [3–4].

Trees establish different symbiotic relationships with a plethora of microorganisms. Therefore, the plant microbiome, also known as a phytobiome, is composed of commensal, mutualistic, and pathogenic microorganisms [5]. The diverse microbial composition contributes to host-microbial homeostasis, meaning the microbial community structure can modulate and maintain tree health over time. It is now known that the colonization of plant tissues by a pathogenic microorganism and disease development may be related to an imbalance of host-microbial homeostasis. In some cases, this imbalance results from atypical environmental conditions or the loss of partner microbes or ancestor microorganisms that are key to tree health.

The microorganisms that are part of the phytobiome include a wide diversity of prokaryotes distributed in different phyla (Proteobacteria, Actinobacteria, Firmicutes, Bacteroidetes, etc.), in addition to fungi (including mycorrhizae), protozoa, nematodes, and viruses. In general, the rhizospheric soil and endosphere are the environments most densely populated by microorganisms; nevertheless, the phylloplane is also an important environment for tree-associated microbial communities. Partner microorganisms can contribute directly and indirectly to tree health through nutrient provision (i.e., phosphorus solubilization and nitrogen fixation), plant hormone production (i.e., gibberellins, auxins), pathogen inhibition via competition, and the inhibition of systemic host resistance.

2. The ecological community of microorganisms

Symbiotic relationships are common to all organisms. Trees establish three main symbiotic relationships with microorganisms: commensalism, mutualism, and parasitism (pathogenic microorganisms). Commensals represent a wide diversity of microorganisms that internally and externally inhabit plant tissues; they benefit from the resources made available by the host tree (contact surface, space, moisture, mucilage, cellular debris, etc.) and do not cause any harm to the host. In contrast, the benefits of commensal microorganisms are considered "neutral" to their host [6]. Although the classical definition of commensalism defines the absence of benefits to the host, this may not be truly applicable to the networks of microbial interactions that develop in plants. The development of a community of commensal microorganisms results in the occupation of a niche that, in their absence, would be held by other groups of microorganisms, including pathogens. Hence, these microorganisms' survival strategies, including cooperation with other commensal microorganisms, pose an obstacle to the pathogen colonization and development in these niches. From this perspective, the benefit of a single commensal species may be neutral, although the gains from the presence of a commensal microbial community are positive to the host, despite not being straightforward to establish the actual gains from the commensal relationship between microorganisms and plants.

While the community of commensal microorganisms can act as an obstacle to establishing primary pathogens, this community can harbor opportunistic pathogens. Factors related to this phenomenon include unusual environmental conditions (i.e., long drought periods or excess water), physical damage to plant tissues (i.e., mechanical damage to the roots, stems, leaves, and fruits), extreme temperatures, and other factors that can affect key components of plants' "innate immunity" [7]. For instance, acute oak decline (AOD) has been a recurring problem in European forests and is associated with opportunistic pathogens, as in the case of the fungus *Armillaria gallica* Marxm. & Romagn (Agaricales: Physalacriaceae). This fungus is an important saprophytic species in wood decay in forests, and it can develop a wide network of hyphae in the subterranean soil and is unable to colonize vigorously growing hosts [8]. Nevertheless, *A. gallica* invades oak trees weakened by insect defoliation or drought, colonizing the root system and causing root rot [9]. The transition from saprophytic to pathogenic lifestyle indicates that a microorganism can alter its relationship with the host depending on environmental conditions and the plant's immune status [9].

The symbiotic mutualistic relationship is a key survival strategy between plants and microorganisms. It is now recognized that mycorrhizal fungi were undoubtedly the most important microorganisms for the successful terrestrial colonization by plants. The evolution of symbiosis with mycorrhizal fungi occurred simultaneously with the establishment of plants on land 450 million years ago [10]. Arbuscular mycorrhizal fungi (AMF—phylum Glomeromycota) were the first fungi to establish a mutualistic symbiosis with plants, and, currently, this group is associated with the roots of over 85% of all plant species [11]. However, the establishment of plants on land also occurred concomitantly with the diversification of other mutualistic symbioses. In forest environments, symbiosis with ectomycorrhizae (ECM—phyla Basidiomycota and Ascomycota) is found in various gymnosperm and angiosperm lineages [12]. Both mycorrhizae groups contribute to the nutrition of their hosts by making scarce nutrients available (i.e., phosphorus and nitrogen), as well as increasing the tolerance of their hosts to abiotic (salinity and water stress) and biotic (pathogen attack) stress. The biogeographical distribution of these symbionts is influenced by climatic factors, such as temperature and rainfall regime. In tropical forests, mutualistic symbiotic association with AMF is more common, whereas ECM fungi are more diverse in temperate and boreal ecosystems [13, 14].

Another important association that may have contributed to the plants' successful terrestrial colonization is symbiosis with nitrogen-fixing bacteria. Association with nitrogen-fixing, nodule-forming bacteria, termed root nodule symbiosis (RNS), is restricted to four angiosperm orders: Fabales, Fagales, Cucurbitales, and Rosales [15]. Most leguminous tree species in tropical forests are capable of forming root nodules and fixing atmospheric nitrogen [16]. In some situations, trees can establish a mutualistic symbiotic association with rhizobia in conjunction with mycorrhizal fungi. The synergistic interaction between these two types of mutualism can improve plant performance, especially in soils nutritionally poor in phosphorus and nitrogen or saline soils. For instance, co-inoculation of the AMF *Rhizophagus fasciculatus* (Thaxt.) Gerd. & Trappe (Glomerales: Glomeraceae) and the actinobacterium Frankia (root nodule symbiont) improved the tolerance to salt stress of the tree species *Casuarina equisetifolia* and *C. obesai* (Fagales) [17].

Unlike RNS, nitrogen fixation through plant-cyanobacteria association is widely distributed in terrestrial plants [18]. Association with nitrogen-fixing cyanobacteria can occur in below-ground (rhizosphere) and above-ground (phyllosphere) environments. These microorganisms can live with mosses growing on the soil, be associated with bryophytes and arboreal epiphytes, or grow in the leaf cavities of plants [18–20]. In fact, evidence has shown that nitrogen fixation by cyanobacteria is associated with "feather moss" (i.e., *Hylocomium splendens*, *Ptilium crista-castrensis*, *Pleurozium schreberi*, and *Sphagnum caprifolium*), an important source of nitrogen input into boreal forest ecosystems [21–22].

The phytobiome also harbors pathogenic microorganisms (symbiotic parasitism relationship). Forest pathogens include fungi, oomycetes, bacteria, phytoplasmas, parasitic higher plants, viruses, and nematodes [23]. Despite a plethora of causative agents, the diseases of forest trees are primarily caused by fungal and oomycete pathogens [24]. The development of some tree diseases has been correlated with changes in the phytobiome. Thus, it is believed that the population growth of a particular pathogen may be related to the decline in the population of beneficial microorganisms. For example, one study reported that pine plants with pine wilt disease (PWD) caused by the nematode *Bursaphelenchus xylophilus* had a lower diversity of beneficial fungi and bacteria in the rhizosphere compared to the rhizosphere of healthy pines [25]. Recent studies have focused on analyzing the microbiome of diseased trees in an attempt to map unknown pathogen groups. Research on the microbiome of oak trees with AOD has reported that this concerning disease is caused by a polymicrobial complex, that is, the onset of AOD is related to the interaction of different bacterial species (*Brenneria goodwinii*, *Gibbsiella quercinecans*, and *Rahnella victoriana*) that work synergistically to develop the disease [26–28]. Some studies suggest that the physiological state of the host can influence the abundance and diversity of pathogenic microorganisms. For example, a recent study evaluated the diversity of foliar endophytes in *Platycladus orientalis* and *Styphnolobium japonicum* trees with different ages (individuals ranging from 10 to 5000 years old for *P. orientalis* and 10 to 1700 years old for *S. japonicum*) [29]. The authors demonstrated that the abundance of latent pathogens (fungi) increased as the trees aged. Thus, the abundance of pathogens *Collectotrichum gloeosporioides* and *Botryosphaeria dothidea* in *S. japonicum* and *Pestalotiopsis funerea* and *Amyloporia subxantha* in *P. orientalis* increased linearly with tree age, indicating that tree age is also an important structuring factor for host communities [29].

Although forest pathology is a recent science, it has been growing rapidly in recent decades mainly due to recent tree death events in forest ecosystems throughout Europe and North America. Forest decline diseases have concerned scientists and government bodies and are becoming increasingly problematic for tree health worldwide [28, 30]. Hence, shedding more light on the role of the tree microbiome will be crucial for properly managing forest environments.

3. Plants as microbial habitats

The plant microbiome is highly dynamic and diverse. Plant-associated microbial communities are deeply influenced by environmental conditions (pH, moisture, temperature, and nutrient availability) and host genotypes. Plants harbor microbial communities in above- and below-ground tissues (**Figure 1**). Below ground, the rhizosphere (soil region in intimate contact with roots) is the environment most densely populated by microorganisms [31]. The root endosphere is another important below-ground region that hosts a vast diversity of microbial communities. The endosphere is the region encompassing the apoplastic spaces in the root cortex (inside the roots). The host genotype strongly influences the microbial communities of the rhizosphere and endosphere, and this can be considered an "extended root phenotype" (i.e., a manifestation of the effects of plant genes on their environment inside and/or outside the organism) [32].

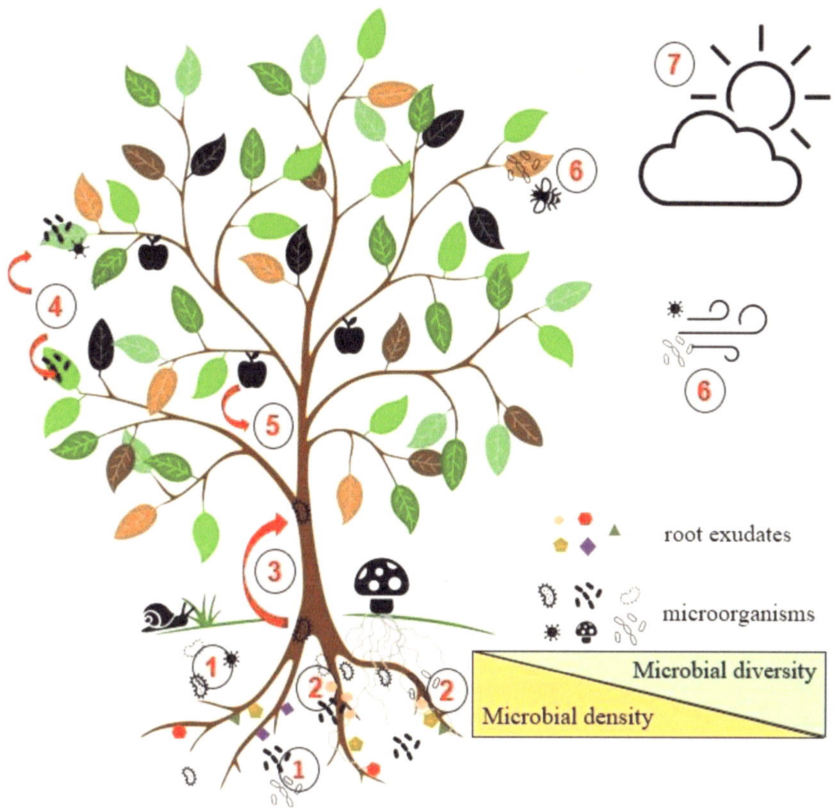

Figure 1.
Interactions between microorganisms and trees. Microorganisms present in the bulk soil that roots can recruit via organic molecule exudation (1). Microorganisms in the rhizosphere and endosphere. The presence of root exudates (sugars, lipids, proteins, and secondary metabolites) stimulates the recruitment of specific microbial taxa. The rhizosphere microbiota is strongly influenced by the physicochemical profile of the exudates of this niche (2). Endophytic microorganisms can inhabit internal regions of the plant without causing disease. Bacteria that inhabit the root endosphere can be translocated via the xylem to other regions of the plant above ground (3). The phyllosphere is the compartment that houses associated microorganisms above ground, and this niche is mainly represented by the leaves (4). Other above-ground compartments may harbor associated microorganisms, such as the outer surface of fruits (carposphere), flowers (anthosphere), and the stem (caulosphere) (5). Microorganisms can be dispersed from one plant to another or from the soil to the phyllosphere by wind, rain, insects, and animals (6). The composition of the microbial community of the phyllosphere is significantly influenced by environmental factors, including solar radiation, temperature, and humidity (7).

Between the volume of soil not occupied by roots and the endosphere region, there is a selection degree of microorganisms. Various studies have demonstrated that microbial species richness is the highest in the bulk soil and that it decreases in the rhizosphere and endosphere (**Figure 1**). In contrast, the population density of specific microorganisms increases from the soil toward the root surface, indicating favorable conditions for the selected microbial species [5]. This phenomenon is called the rhizospheric effect, that is, the composition of root exudates (sugars, oligosaccharides, vitamins, nucleotides, flavones, auxins, and secondary metabolites) modulates the physicochemical conditions of the rhizosphere region and, thus, the plant can recruit and select groups of microorganisms that can proliferate in the specific physicochemical conditions of the root zone [33]. Estimates have indicated that up to 40% of the carbon reserves fixed by photosynthesis are provided in the rhizosphere, indicating the active role of plants in recruiting microbial communities. Although the root zone harbors a wealth of biodiversity, the rhizosphere and endosphere microbial communities are dominated by four bacterial phyla: Actinobacteria, Bacteroidetes, Firmicutes, and Proteobacteria [5, 34]. In the bulk forest soil, the dominant fungal group is Basidiomycota, while Ascomycota is the most prevalent group in plant tissues [9].

Evidence suggests that plants adapt to biotic stress by altering their root exudate chemistry to assemble health-promoting microbiomes [35]. This phenomenon is termed the "cry-for-help" hypothesis, and it posits that plants modulate the chemistry of root exudates to recruit partner microorganisms capable of increasing the plants tolerance to a given stress condition, such as insect herbivory, pathogen attack, or nutrient shortage. Therefore, the chemical composition of root exudates influences the metabolism of rhizosphere microbial taxa while the recruited microbiota assists plant homeostasis by encoding functionalities that extend the plant genome [36]. For example, a study with the model plant *Arabidopsis thaliana* and inoculated with the Gram-negative pathogen *Pseudomonas syringae* demonstrated that subsequent generations of *A. thaliana* subjected to inoculation with the pathogen were able to modulate the root exudation profile, alter the composition of the rhizosphere microbial community, and increase the disease suppressive response [37].

Above ground, there is also an important plant compartment that hosts microorganisms; it is called the phyllosphere and refers mainly to the leaves. However, there are also other important plant sub-compartments above ground, such as the anthosphere (external environment of flowers), caulosphere (environment of the plant stem), carposphere (external surface of fruits), and spicosphere (niche formed in plants with spikes) [38–39]. The phyllosphere is an oligotrophic environment subject to severe modifications in a short period of time (temperature, humidity, and radiation fluctuations); despite being disconnected from the soil, this environment indirectly influences the phyllosphere. Dust particles from the soil can be dispersed by the wind and deposited in the above-ground plant compartments and thus provide nutrients for the microbial communities of the phyllosphere. In addition, microorganisms can be dispersed from the soil to the above-ground part of the plant by wind or colonize the phyllosphere after being recruited in the rhizosphere and systematically translocated to the above-ground part via the xylem. Microorganisms that can colonize plant tissues internally or translocate via the xylem to different tissues of the plant above ground without causing disease are called endophytes. These microorganisms may live part of their life cycle associated with the root endosphere region or translocated to plant tissues above ground. Furthermore, as in the rhizosphere, the bacterial communities of the phyllosphere are dominated by the taxa Actinobacteria, Bacteroidetes, Firmicutes, and Proteobacteria [40].

The phyllosphere is an open environment where the arrival and departure of taxa are constant [41]. The phyllosphere generally consists of microorganisms that are dispersed mainly from the soil and reach the surface of leaves and stem through rain, wind, and insects (**Figure 1**). The survivability and colonization of these microorganisms depend on several factors, such as host genotype and health (since both modulate the chemical composition of leaves), water availability, temperature fluctuation, UV exposure, chemical applications (i.e., fertilizers and pesticides), and inter- and intraspecific microbial competition [41–43]. Although several factors influence microbial communities, it is recognized that environmental conditions are the key factors that determine population structure in microbial communities of the phyllosphere. For instance, a recent study revealed that the microbial community composition of the phyllosphere of spruce trees was influenced by seasonal changes, whereas the bacterial and fungal communities of the rhizosphere of these same trees were influenced by anthropogenic nutrient availability [44].

Because it is an environment with harsh conditions, many microorganisms have evolved and developed survival strategies to colonize the phyllosphere. Thus, to colonize niches in the phyllosphere, some microorganisms can alter the chemistry of the leaf surface or use specialized cellular structures to increase their ability to compete. For example, the bacterium *P. syringae* (an important plant pathogen) releases surfactants that increase mobility and local water availability on leaf surfaces. In addition, this microorganism has flagella that favor bacterial motility on the leaf surface. In the case of *P. syringae*, these strategies are related to its virulence [41]. Nonetheless, characteristics of the host plant's aerial part structures also influence the successful colonization of the phyllosphere. One study evaluated the bacterial functional diversity in the phyllosphere of different tree species in a Neotropical forest and found evidence suggesting an adaptive correspondence between phyllosphere microorganisms and their tree hosts [40]. The authors demonstrated that tree characteristics, such as leaf morphology and leaf metal contents (copper, manganese, and zinc), are correlated with phyllosphere microbial community composition [40].

4. Functions of the host-associated microbiota

Plant-associated microorganisms can assist the health of the host in different ways. Many endophytic microorganisms can make previously scarce nutrients available in the soil or increase host tolerance to stressors. Mycorrhizae play an important role in maintaining tree health among endophytic microorganisms. The role of these microorganisms is related not only to making phosphorus and nitrogen available to the host but also to increasing plant tolerance to water stress and pathogen attack. In one experiment, the authors evaluated the effects of inoculation of an AMF (*Glomus etunicatum*) on pistachio (*Pistacia vera*) seedlings subjected to water stress. They observed that plants inoculated with the fungus had an increased tolerance to water stress compared to the control (seedlings subjected to water stress and not inoculated) [45]. Among the mechanisms related to increased stress tolerance and possibly induced by the mycorrhizal association, the authors highlighted a greater accumulation of osmotic adjustment compounds (i.e., soluble sugar content), increased activity of antioxidant enzymes (i.e., catalase and peroxidase), secondary metabolite production (i.e., flavonoids), and nutrient accumulation (nitrogen and calcium) [45].

Further findings have also suggested that association with mycorrhizal fungi improves host resistance to pathogens [46]. For instance, a recent study used amplicon sequencing to determine the presence of a wide taxonomic range in the rhizosphere of apple (*Malus domestica* Borkh) rootstocks [47]. The authors observed that roots of the G.890 rootstock (which is tolerant to apple replant disease—ARD) harbored a significantly higher percentage of AMF species, indicating a possible active role of endophytic fungal communities in apple tree tolerance to soil pathogens (including *Rhizoctonia* spp., *Phytophthora* spp., and *Pratylenchus penetrans*) that cause ARD [47]. In another study [25], the authors evaluated the effects of inoculating ECM species (*Suillus bovinus* and *Amanita vaginata*) and dark septate endophytes (DSE, *Gaeumannomyces cylindrosporus* and *Paraphoma chrysanthemicola*) on the tolerance of pines (*Pinus tabulaeformis*) to PWD caused by the pine nematode *Bursaphelenchus xylophilus*. The authors demonstrated that inoculating pines with ECM/DSE reduced disease severity caused by *B. xylophilus* and increased the recruitment of beneficial bacterial and fungal groups in the rhizosphere of pines [25]. In addition to improving nutrition and stimulating defense mechanisms in their hosts, ectomycorrhizal fungi can assist their host against pathogen attacks by forming a thick fungal mantle that acts as a mechanical barrier against the penetration of soil pathogens [48].

Ectomycorrhizae are known to benefit their host trees in different ways. This mutualistic symbiosis is especially useful for forest plantations intended for the production of wood, cellulose, or the recovery of degraded areas. However, mutualism between trees and ECM species can stimulate another highly profitable economic activity: truffle farming. Truffles are the reproductive structures of hypogean ectomycorrhizal fungi and are appreciated in haute cuisine [49]. A recent study evaluated the effect of mycorrhization of pecan trees under subtropical conditions in Brazil [50]. The authors demonstrated that inoculation of pecan seedlings with the ectomycorrhizal species *Tuber aestivum* and *T. brumale* increased plant growth and, in addition, ECM produced high-value edible structures (truffles) [50]. These results indicate the possibility of developing a highly profitable economic activity associated with pecan orchards in subtropical regions.

Recruitment of plant growth-promoting bacteria (PGPB) is another important strategy trees adopt to increase their resistance to pathogen attacks; PGPB is known to promote plant growth through different mechanisms, including producing or stimulating plant hormones (i.e., gibberellins, auxins, and cytokinins), nitrogen fixation, phosphate solubilization, among others. In addition, PGPB may exhibit biocontrol activity against plant pathogens; numerous mechanisms have been described as being responsible for the biocontrol ability of PGPB, such as direct competition for space, emission of volatile compounds, siderophore production, lytic enzymes (i.e., proteases, chitinases, and lipases), antibiotics (i.e., amphisin, 2,4-diacetylphloroglucinol, and oomycin A), and induction of systemic resistance of the host plant [51–53].

The ability of the plant to recruit partner microorganisms indicates that the rhizosphere is a reservoir of beneficial microorganisms amenable to selecting and applying disease biocontrol programs. For example, an endophytic PGPB (*Bacillus velezensis* OEE1) was isolated from olive (*Olea europaea*) roots, and its disease biocontrol ability was tested against fungal (*Fusarium solani*, *Botrytis cinera*, etc.) and oomycete (*Phytophthora* spp.) pathogens [54]. The authors observed that the biocontrol ability of fungal pathogens and oomycetes by the isolate *B. velezensis* OEE1 was related to a wide range of competitive characteristics, including phosphate solubilization and producing siderophores, extracellular hydrolytic enzymes (amylases, cellulases, and pectinases), biosurfactant (surfactins), and secondary metabolites [54]. In another study, bacteria were isolated from the rhizosphere of avocado trees that survived root rot infestations caused by the oomycete *Phytophthora cinnamomi* [55]. The authors evaluated the antagonistic activity of the isolates against *P. cinnamomi* and selected a potential PGPB (*Bacillus acidiceler*) that could inhibit the growth of the oomycete by producing volatile compounds, indicating the potential use of this PGPB in biocontrol programs for pathogenic oomycetes [55].

5. Conclusions

The microbial communities of the rhizosphere and phyllosphere are essential to the health of their hosts. Researchers are currently seeking to understand the active role of tree-associated microorganisms and the mechanisms related to the recruitment of beneficial taxa. Knowledge about tree-associated microorganisms is increasingly more important in the face of the current scenario of increasing diseases of forest declines. Therefore, microbial ecology studies applied to tree-associated bacteria and fungi communities will enable researchers to bioprospect new microorganisms for use in plant growth promotion and disease control.

Acknowledgements

The authors are grateful to the National Council of Scientific and Technological Development (Conselho Nacional de Desenvolvimento Científico e Tecnológico—CNPq) and Coordination for the Improvement of Higher Education Personnel (Coordenação de Aperfeiçoamento de Pessoal de Nível Superior—CAPES) (Finance Code 001) for providing scholarships and financial support for this research. We would also like to thank Atlas Assessoria Linguística for language editing.

Conflict of interest

The authors declare no conflict of interest.

Author details

Rodrigo F. Ramos[1], Lisiane Sobucki[2], Estéfany Pawlowski[2], Janaina S. Sarzi[2], Jessica E. Rabuske[3], Lucas G. Savian[2], Tiago E. Kaspary[4] and Cristiano Bellé[1]*

1 Instituto Phytus - Staphyt Brasil, Itaara, Brazil

2 Federal University of Santa Maria, Santa Maria, Brazil

3 Regional Integrated University of Upper Uruguai and Missions, Frederico Westphalen, Brazil

4 National Institute of Agricultural Research – INIA La Estanzuela, Colonia del Sacramento, Uruguay

*Address all correspondence to: crbelle@gmail.com

References

[1] Rosenberg E, Zilber-Rosenberg I. The hologenome concept of evolution after 10 years. Microbiome. 2018;**6**:78

[2] Hassani MA, Dúran P, Hacquard S. Microbial interactions within the plant holobiont. Microbiome. 2018;**6**:58

[3] Cregger MA, Veach AM, Yang ZK, Crouch MJ, Vilgalys R, Tuskan GA, et al. The *Populus* holobiont: Dissecting the effects of plant niches and genotype on the microbiome. Microbiome. 2018;**6**:31

[4] Vandenkoornhuyse P, Quaiser A, Duhamel M, Le Van A, Dufresne A. The importance of the microbiome of the plant holobiont. The New Phytologist. 2015;**206**:1196-1206

[5] Rodriguez PA, Rothballer M, Chowdhury SP, Nussbaumer T, Gutjahr C, Falter-Braun P. Systems biology of plant-microbiome interactions. Molecular Plant. 2019;**12**:804-821

[6] Terhonen E, Blumenstein K, Kovalchuk A, Asiegbu FO. Forest tree microbiomes and associated fungal endophytes: Functional roles and impact on forest health. Forests. 2019;**10**:42

[7] Charkowski AO. Opportunistic pathogens of terrestrial plants. In: Hurst CJ, editor. The Rasputin Effect: When Commensals and Symbionts Become Parasitic. Advances in Environmental Microbiology, Vol 3. Cham: Springer; 2016. p. 147-168

[8] Lee CA, Holdo RM, Muzika R-M. Feedbacks between forest structure and an opportunistic fungal pathogen. Journal of Ecology. 2021;**109**:4092-4102

[9] Marçais B, Bréda N. Role of an opportunistic pathogen in the decline of stressed oak trees. Journal of Ecology. 2006;**94**:1214-1223

[10] Brundrett MC, Tedersoo L. Evolutionary history of mycorrhizal symbioses and global host plant diversity. The New Phytologist. 2018;**220**:1108-1115

[11] Strullu-Derrien C, Selosse MA, Kenrick P, Martin FM. The origin and evolution of mycorrhizal symbioses: From palaeomycology to phylogenomics. The New Phytologist. 2018;**220**:1012-1030

[12] Radhakrishnan GV, Keller J, Rich MK, Vernié T, Mbadinga DLM, Vigneron N, et al. An ancestral signalling pathway is conserved in intracellular symbioses-forming plant lineages. Nature Plants. 2020;**6**:280-289

[13] Corrales A, Henkel TW, Smith ME. Ectomycorrhizal associations in the tropics - biogeography, diversity patterns and ecosystem roles. The New Phytologist. 2018;**220**:1076-1091

[14] Tedersoo L, Bahram M, Põlme S, Kõljalg U, Yorou NS, Wijesundera R, et al. Global diversity and geography of soil fungi. Science. 2014;**346**:1052

[15] Griesmann M, Chang Y, Liu X, Song Y, Haberer G, Crook MB, et al. Phylogenomics reveals multiple losses of nitrogen-fixing root nodule symbiosis. Science. 2018;**361**:eaat1743

[16] Chaer GM, Resende AS, Campello EFC, Faria SM, Boddey RM. Nitrogen-fixing legume tree species for the reclamation of severely degraded lands in Brazil. Tree Physiology. 2011;**31**:139-149

[17] Djighaly PI, Diagne N, Ngom D, Cooper K, Pignoly S, Hocher JMF, et al. Effect of symbiotic associations with Frankia and arbuscular mycorrhizal fungi on antioxidant activity and cell ultrastructure in *C. equisetifolia* and *C. obesa* under salt stress. Journal of Forest Research. 2022;**27**(2):117-127

[18] Delaux P-M, Radhakrishnan G, Oldroyd G. Tracing the evolutionary path to nitrogen-fixing crops. Current Opinion in Plant Biology. 2015;**26**:95-99

[19] DeLuca TH, Zackrisson O, Nilsson MC, Sun S, Crespo MA. Long-term fate of nitrogen fixation in *Pleurozium schreberi* Brid (Mit.) moss carpets in boreal forests. Applied Soil Ecology. 2022;**169**:104215

[20] Markham J, Otárola MF. Bryophyte and lichen biomass and nitrogen fixation in a high elevation cloud forest in Cerro de La Muerte. Costa Rica. Oecologia. 2021;**195**:489-497

[21] Renaudin M, Blasi C, Bradley RL, Bellenger JF. New insights into the drivers of moss-associated nitrogen fixation and cyanobacterial biomass in the eastern Canadian boreal forest. Journal of Ecology. 2022;**110**:1403-1418

[22] Ininbergs K, Bay G, Rasmussen U, Wardle DA, Nilsson MC. Composition and diversity of *nifH* genes of nitrogen-fixing cyanobacteria associated with boreal forest feather mosses. The New Phytologist. 2011;**192**:507-517

[23] Sturrock RN, Frankel SJ, Brown AV, Hennon PE, Kliejunas JT, Lewis KJ, et al. Climate change and forest diseases. Plant Pathology. 2011;**60**:133-149

[24] Desprez-Loustau M-L, Aguayo J, Dutech C, Hayden KJ, Husson C, Jakushkin B, et al. An evolutionary ecology perspective to address forest pathology challenges of today and tomorrow. Annals of Forest Science. 2016;**73**:45-67

[25] Chu H, Wang H, Zhang Y, Li Y, Wang C, Dai D, et al. Inoculation with ectomycorrhizal fungi and dark septate endophytes contributes to the resistance of *Pinus* spp. to pine wilt disease. Frontiers in Microbiology. 2021;**12**:687304

[26] Doonan JM, Broberg M, Denman S, McDonald JE. Host–microbiota–insect interactions drive emergent virulence in a complex tree disease. Proceedings of the Royal Society B: Biological Sciences. 2020;**287**:20200956

[27] Ruffner B, Schneider S, Meyer JB, Queloz V, Rigling D. First report of acute oak decline disease of native and non-native oaks in Switzerland. New Disease Reports. 2020;**41**:18

[28] Denman S, Doonan J, Ransom-Jones E, Broberg M, Plummer S, Kirk S, et al. Microbiome and infectivity studies reveal complex polyspecies tree disease in acute oak decline. The ISME Journal. 2018;**12**:386-399

[29] Yu Z, Ding H, Shen K, Bu F, Newcombe G, Liu H. Foliar endophytes in trees varying greatly in age. European Journal of Plant Pathology. 2021;**160**:375-384

[30] Pinho D, Barroso C, Froufe H, Brown N, Vanguelova E, Egas C, et al. Linking tree health, rhizosphere physicochemical properties, and microbiome in acute oak decline. Forests. 2020;**11**:1153

[31] York LM, Carminati A, Mooney SJ, Ritz K, Bennett MJ. The holistic rhizosphere: Integrating zones, processes, and semantics in the soil influenced by roots. Journal of Experimental Botany. 2016;**67**:3629-3643

[32] Cantó CF, Simonin M, King K, Moulin L, Bennett MJ, Castrillo G, et al. An extended root phenotype: the rhizosphere, its formation and impacts on plant fitness. The Plant Journal. 2020;**103**:951-964

[33] Dotaniya ML, Meena VD. Rhizosphere effect on nutrient availability in soil and its uptake by plants: A review. Proceedings of the National Academy of Sciences, India Section B: Biological Sciences. 2015;**85**:1-12

[34] Soussi A, Ferjani R, Marasco R, Guesmi A, Cherif H, Rolli E, et al. Plant-associated microbiomes in arid lands: Diversity, ecology and biotechnological potential. Plant and Soil. 2016;**405**:357-370

[35] Rolfe SA, Griffiths J, Ton J. Crying out for help with root exudates: Adaptive mechanisms by which stressed plants assemble health-promoting soil microbiomes. Current Opinion in Microbiology. 2019;**49**:73-82

[36] Rolli E, Vergani L, Ghitti E, Patania G, Mapelli F, Borin S. 'Cry-for-help' in contaminated soil: A dialogue among plants and soil microbiome to survive in hostile conditions. Environmental Microbiology. 2021;**23**:5690-5703

[37] Yuan J, Zhao J, Wen T, Zhao M, Li R, Goossens P, et al. Root exudates drive the soil-borne legacy of aboveground pathogen infection. Microbiome. 2018;**6**:156

[38] Kavamura VN, Mendes R, Bargaz A, Mauchline TH. Defining the wheat microbiome: Towards microbiome-facilitated crop production. Computational and Structural Biotechnology Journal. 2021;**19**:1200-1213

[39] Compant S, Samad A, Faist H, Sessitsch A. A review on the plant microbiome: Ecology, functions, and emerging trends in microbial application. Journal of Advanced Research. 2019;**19**:29-37

[40] Lajoie G, Maglione R, Kembel SW. Adaptive matching between phyllosphere bacteria and their tree hosts in a neotropical forest. Microbiome. 2020;**8**:1-10

[41] Koskella B. The phyllosphere. Current Biology. 2020;**30**:R1143

[42] Liu H, Brettell LE, Singh B. Linking the phyllosphere microbiome to plant health. Trends in Plant Science. 2020;**25**:841-844

[43] Vorholt JA. Microbial life in the phyllosphere. Nature Reviews. Microbiology. 2012;**10**:828-840

[44] Haas JC, Street NR, Sjödin A, Lee NM, Högberg MN, Näsholm T, et al. Microbial community response to growing season and plant nutrient optimisation in a boreal Norway spruce forest. Soil Biology and Biochemistry. 2018;**125**:197-209

[45] Abbaspour H, Saeidi-Sar S, Afshari H, Abdel-Wahhab MA. Tolerance of mycorrhiza infected Pistachio (*Pistacia vera* L.) seedling to drought stress under glasshouse conditions. Journal of Plant Physiology. 2012;**169**:704-709

[46] Cowan JA, Gehring CA, Ilstedt U, Grady KC. Host identity and neighborhood trees affect belowground microbial communities in a tropical rainforest. Tropical Ecology. 2021;**63**:216-228

[47] Van Horn C, Somera TS, Mazzola M. Comparative analysis of the rhizosphere and endophytic microbiomes

across apple rootstock genotypes in replant orchard soils. Phytobiomes. 2021;**5**:231-243

[48] Khullar S, Reddy MS. Ectomycorrhizal Diversity and Tree Sustainability. In: Satyanarayana T, Das SK, Johri BN, editors. Microbial Diversity in Ecosystem Sustainability and Biotechnological Applications. Singapore: Springer; 2019. pp. 145-166

[49] Sulzbacher MA, Hamann JJ, Fronza D, Jacques RJS, Giachini AJ, Grebenc G, et al. Ectomycorrhizal fungi in pecan orchards and the potential of truffle cultivation in Brazil. Ciência Florestal. 2019;**29**:975-987

[50] Freiberg JA, Sulzbacher MA, Grebenc T, Santana NA, Schardong IS, Mazorri G, et al. Mycorrhization of pecans with European truffles (*Tuber* spp., Tuberaceae) under southern subtropical conditions. Applied Soil Ecology. 2021;**168**:104108

[51] Orozco-Mosqueda MC, Rocha-Granados MC, Glick BR, Santoyo G. Microbiome engineering to improve biocontrol and plant growth-promoting mechanisms. Microbiological Research. 2018;**208**:25-31

[52] Olanrewaju OS, Glick BR, Babalola OO. Mechanisms of action of plant growth promoting bacteria. World Journal of Microbiology and Biotechnology. 2017;**33**:197

[53] Compant S, Duffy B, Nowak J, Clément C, Barka EA. Use of plant growth-promoting bacteria for biocontrol of plant diseases: Principles, mechanisms of action, and future prospects. Applied and Environmental Microbiology. 2005;**71**:4951-4959

[54] Cheffi M, Bouket AC, Alenezi FN, Luptakova L, Belka M, Vallat A, et al. Root endophyte *Bacillus velezensis* OEE1 counteracts oomycete and fungal harmful pathogens and harbours a large repertoire of secreted and volatile metabolites and beneficial functional genes. Microorganisms. 2019;**7**:314

[55] Méndez-Bravo A, Cortazar-Murillo EM, Guevara-Avendaño E, Ceballos-Luna O, Rodríguez-Haas B, Kiel-Martínez AL, et al. Plant growth-promoting rhizobacteria associated with avocado display antagonistic activity against *Phytophthora cinnamomi* through volatile emissions. PLoS One. 2018;**13**:e0194665